METHODS IN MOLECULAR BIOLOGY

Series Editor
John M. Walker
School of Life and Medical Sciences
University of Hertfordshire
Hatfield, Hertfordshire, AL10 9AB, UK

For further volumes:
http://www.springer.com/series/7651

Microchip Diagnostics

Methods and Protocols

Edited by

Valérie Taly

INSERM UMRS1147, CNRS SNC 5014, Université Paris Descartes, Equipe labellisée Ligue Nationale contre le cancer 2016, Paris, France

Jean-Louis Viovy

Macromolecules and Microsystems in Biology and Medicine, Institut Curie, Centre National de la Recherche Scientifique, Université Pierre et Marie Curie, PSL Research University, UMR 168, Paris, France

Stéphanie Descroix

Macromolecules and Microsystems in Biology and Medicine, Institut Curie, Centre National de la Recherche Scientifique, Université Pierre et Marie Curie, PSL Research University, UMR 168, Paris, France

Humana Press

Editors
Valérie Taly
INSERM UMRS1147, CNRS SNC 5014
Université Paris Descartes, Equipe labellisée Ligue Nationale contre le cancer 2016
Paris, France

Stéphanie Descroix
Macromolecules and Microsystems in Biology and Medicine
Institut Curie, Centre National de la Recherche Scientifique, Université Pierre et Marie Curie, PSL Research University, UMR 168
Paris, France

Jean-Louis Viovy
Macromolecules and Microsystems in Biology and Medicine
Institut Curie, Centre National de la Recherche Scientifique, Université Pierre et Marie Curie, PSL Research University, UMR 168
Paris, France

ISSN 1064-3745 ISSN 1940-6029 (electronic)
Methods in Molecular Biology
ISBN 978-1-4939-6732-2 ISBN 978-1-4939-6734-6 (eBook)
DOI 10.1007/978-1-4939-6734-6

Library of Congress Control Number: 2016963676

Printed on acid-free paper

This Humana Press imprint is published by Springer Nature
The registered company is Springer Science+Business Media LLC
The registered company address is: 233 Spring Street, New York, NY 10013, U.S.A.

Preface

By allowing unprecedented level of miniaturization, parallelization, and integration, microfluidics is increasingly becoming a method of choice to develop new diagnostic tools. Recent advances in microfluidics technology have the potential to revolutionize the field of diagnosis. Researchers have been developing new tools and strategies allowing the highly sensitive detection and fine characterization of biomarkers using cutting-edge developments. However, few of these works really meet the necessary requirements allowing their clear validation as diagnostic tools. This book presents an overview of recent advances in microfluidics for diagnostic purposes with a specific focus on the application side and the pertinence of the developed devices and procedures for improving human health. The aim of this book is to illustrate how microfluidic approaches can meet the requirement of clinical diagnosis based on molecular or cellular biomarkers. As microfluidics has reached an unprecedented level of maturity, the first part of this book is dedicated to the business aspects of microfluidic-based systems for diagnosis. The second part gathers contributions highlighting how versatile microfluidics can be regarding protein bioassay integration. Microfluidic approaches for nucleic analysis based on mono- or diphasic format are reported in the third part. The last part brings together contributions on a more recent field of applications in microfluidics: cell analysis and enrichment.

Paris, France

Valérie Taly
Jean-Louis Viovy
Stéphanie Descroix

Contents

Preface . . . . *v*
Contributors . . . . *ix*

PART I

1 Microfluidics-Enabled Diagnostic Systems: Markets, Challenges, and Examples 3
Holger Becker and Claudia Gärtner

PART II MICROCHIPS FOR PROTEIN BIOASSAYS

2 Capillary-Driven Microfluidic Chips for Miniaturized Immunoassays: Efficient Fabrication and Sealing of Chips Using a "Chip-Olate" Process. 25
Yuksel Temiz and Emmanuel Delamarche

3 Capillary-Driven Microfluidic Chips for Miniaturized Immunoassays: Patterning Capture Antibodies Using Microcontact Printing and Dry-Film Resists. 37
Yuksel Temiz, Robert D. Lovchik, and Emmanuel Delamarche

4 Fabrication and Evaluation of Microfluidic Immunoassay Devices with Antibody-Immobilized Microbeads Retained in Porous Hydrogel Micropillars. 49
Toshihiro Kasama, Noritada Kaji, Manabu Tokeshi, and Yoshinobu Baba

5 Using Electrophoretic Immunoassay to Monitor Hormone Secretion 57
Shusheng Lu and Robert T Kennedy

6 On-Chip Magnetic Particle-Based Immunoassays Using Multilaminar Flow for Clinical Diagnostics 69
Mark D. Tarn and Nicole Pamme

7 Digital Microfluidics Assisted Sealing of Individual Magnetic Particles in Femtoliter-Sized Reaction Wells for Single-Molecule Detection. 85
Deborah Decrop, Elena Pérez Ruiz, Phalguni Tewari Kumar, Lisa Tripodi, Tadej Kokalj, and Jeroen Lammertyn

PART III MICROCHIPS FOR NUCLEIC ACIDS ANALYSIS

8 Microfluidic DNA Stretching Device for Single-Molecule Diagnostics. 105
Daisuke Onoshima and Yoshinobu Baba

9 Droplet Microfluidic and Magnetic Particles Platform for Cancer Typing 113
Davide Ferraro, Jérôme Champ, Bruno Teste, M. Serra, Laurent Malaquin, Stéphanie Descroix, Patricia de Cremoux, and Jean-Louis Viovy

10 SlipChip Device for Digital Nucleic Acid Amplification 123
Feng Shen

11 Multiplex Detection of KRAS Mutations Using Passive Droplet Fusion...... 133
Deniz Pekin and Valerie Taly

12 Droplet-Based Microfluidics Digital PCR for the Detection of KRAS Mutations.................................... 143
Deniz Pekin and Valerie Taly

PART IV MICROFLUIDIC CHIPS FOR CELL-BASED ENRICHMENT AND ANALYSIS

13 Negative Enrichment of Circulating Tumor Cells in Blood Using a Microfluidic Chip.................................... 167
Hamizah A. Cognart and Chia-Pin Chang

14 Microfluidic-Based Bacteria Isolation from Whole Blood for Diagnostics of Blood Stream Infection............................ 175
Sergey Zelenin, Harisha Ramachandraiah, Asim Faridi, and Aman Russom

15 Microfluidic Platform for Parallel Single Cell Analysis for Diagnostic Applications.................................... 187
Séverine Le Gac

16 FISH-in-CHIPS: A Microfluidic Platform for Molecular Typing of Cancer Cells.................................... 211
Karla Perez-Toralla, Guillaume Mottet, Ezgi Tulukcuoglu-Guneri, Jérôme Champ, François-Clément Bidard, Jean-Yves Pierga, Jerzy Klijanienko, Irena Draskovic, Laurent Malaquin, Jean-Louis Viovy, and Stéphanie Descroix

Index *221*

Contributors

YOSHINOBU BABA • *Department of Applied Chemistry, Graduate School of Engineering, Nagoya University, Japan; ImPACT Research Center for Advanced Nanobiodevices, Nagoya University, Nagoya, Japan; Division of Biotechnology and Macromolecular Chemistry, Hokkaido University, Sapporo, Japan; Institute of Innovation for Future Society, Nagoya University, Nagoya, Japan; Health Research Institute, National Institute of Advanced Industrial Science and Technology, (AIST), Takamatsu, Japan*

HOLGER BECKER • *Microfluidic ChipShop GmbH, Jena, Germany*

FRANÇOIS-CLÉMENT BIDARD • *Department of Medical Oncology, Institut Curie, Paris, France*

JÉRÔME CHAMP • *Macromolecules and Microsystems in Biology and Medicine, Institut Curie, Centre National de la Recherche Scientifique, Université Pierre et Marie Curie, PSL Research University, UMR 168, Paris, France; University Paris-Diderot, PRES Paris Cité, INSERM/CNRS UMR944/7212, Paris, France; Molecular Oncology Unit, APHP, Saint-Louis Hospital, Paris, France*

CHIA-PIN CHANG • *BioElectronics Programme, Institute of Microelectronics, A*STAR (Agency for Science, Technology and Research), Singapore, Singapore*

HAMIZAH A. COGNART • *Institut Curie, PSL Research University, CNRS, UMR 168, Paris, France; Sorbonne Universités, UPMC University Paris 06, CNRS, UMR 168, Paris, France*

PATRICIA DE CREMOUX • *University Paris-Diderot, PRES Paris Cité, INSERM/CNRS UMR944/7212, Paris, France; Molecular Oncology Unit, APHP, Saint-Louis Hospital, Paris, France*

DEBORAH DECROP • *MeBioS - Biosensors, Department of Biosystems, KU Leuven, Leuven, Belgium*

EMMANUEL DELAMARCHE • *IBM Research GmbH, Rüschlikon, Switzerland*

STÉPHANIE DESCROIX • *Macromolecules and Microsystems in Biology and Medicine, Institut Curie, Centre National de la Recherche Scientifique, Université Pierre et Marie Curie, PSL Research University, UMR 168, Paris, France*

IRENA DRASKOVIC • *Telomeres and Cancer laboratory, Institut Curie, UPMC Univ. Paris 06, Equipe labellisée «Ligue», Paris, France*

ASIM FARIDI • *Science for Life Laboratory, Division of Proteomics and Nanobiotechnology, KTH Royal Institute of Technology, Stockholm, Sweden*

DAVIDE FERRARO • *Macromolecules and Microsystems in Biology and Medicine, Institut Curie, Centre National de la Recherche Scientifique, Université Pierre et Marie Curie, PSL Research University, UMR 168, Paris, France*

SÉVERINE LE GAC • *Applied Microfluidics for BioEngineering Research, MESA+ Institute for Nanotechnology and MIRA Institute for Biomedical Engineering and Technical Medicine, University of Twente, Enschede, The Netherlands*

CLAUDIA GÄRTNER • *Microfluidic ChipShop GmbH, Jena, Germany*

EZGI TULUKCUOGLU-GUNERI • *Macromolecules and Microsystems in Biology and Medicine, Institut Curie, Centre National de la Recherche Scientifique, Université Pierre et Marie Curie, PSL Research University, UMR 168, Paris, France*

Noritada Kaji • *Department of Applied Chemistry, Graduate School of Engineering, Nagoya University, Nagoya, Japan; ImPACT Research Center for Advanced Nanobiodevices, Nagoya University, Furo-cho, Chikusa-ku, Nagoya, Japan*

Toshihiro Kasama • *Department of Applied Chemistry, Graduate School of Engineering, Nagoya University, Nagoya, Japan*

Robert T. Kennedy • *Department of Chemistry, University of Michigan, Ann Arbor, Michigan, USA; Department of Pharmacology, University of Michigan, Ann Arbor, Michigan, USA*

Jerzy Klijanienko • *Department of Pathology, Institut Curie, Paris, France*

Tadej Kokalj • *MeBioS - Biosensors, Department of Biosystems, KU Leuven, Leuven, Belgium*

Phalguni Tewari Kumar • *MeBioS - Biosensors, Department of Biosystems, KU Leuven, Leuven, Belgium*

Jeroen Lammertyn • *MeBioS - Biosensors, Department of Biosystems, KU Leuven, Leuven, Belgium*

Robert D. Lovchik • *IBM Research GmbH, Rüschlikon, Switzerland*

Shusheng Lu • *Department of Chemistry, University of Michigan, Ann Arbor, MI, USA*

Laurent Malaquin • *Macromolecules and Microsystems in Biology and Medicine, Institut Curie, Centre National de la Recherche Scientifique, Université Pierre et Marie Curie, PSL Research University, UMR 168, Paris, France*

Guillaume Mottet • *Macromolecules and Microsystems in Biology and Medicine, Institut Curie, Centre National de la Recherche Scientifique, Université Pierre et Marie Curie, PSL Research University, UMR 168, Paris, France*

Daisuke Onoshima • *Institute of Innovation for Future Society, Nagoya University, Nagoya, Japan; ImPACT Research Center for Advanced Nanobiodevices, Nagoya University, Nagoya, Japan*

Nicole Pamme • *Department of Chemistry, The University of Hull, Hull, UK*

Deniz Pekin • *CNRS, Univ. Bordeaux, CRPP, UPR 8641, Pessac, France*

Karla Perez-Toralla • *Macromolecules and Microsystems in Biology and Medicine, Institut Curie, Centre National de la Recherche Scientifique, Université Pierre et Marie Curie, PSL Research University, UMR 168, Paris, France; Univ Paris Diderot, Sorbonne Paris Cité, Paris, France; Université Paris Sorbonne Cité, INSERM UMR-S1147, CNRS SNC 5014, Equipe labellisée Ligue Nationale contre le cancer, Centre Universitaire des Saints-Pères, Paris, France*

Jean-Yves Pierga • *Department of Medical Oncology, Institut Curie, Paris, France; Université Paris Descartes, Paris, France*

Harisha Ramachandraiah • *Science for Life Laboratory, Division of Proteomics and Nanobiotechnology, KTH Royal Institute of Technology, Stockholm, Sweden*

Elena Pérez Ruiz • *MeBioS - Biosensors, Department of Biosystems, KU Leuven, Leuven, Belgium*

Aman Russom • *Science for Life Laboratory, Division of Proteomics and Nanobiotechnology, KTH Royal Institute of Technology, Stockholm, Sweden*

M. Serra • *Macromolecules and Microsystems in Biology and Medicine, Institut Curie, Centre National de la Recherche Scientifique, Université Pierre et Marie Curie, PSL Research University, UMR 168, Paris, France*

Feng Shen • *SlipChip Corporation, San Jose, CA, USA*

Valerie Taly • *Université Paris Sorbonne Cité, INSERM UMR-S1147, CNRS SNC 5014, Equipe labellisée Ligue Nationale contre le cancer, Centre Universitaire des Saints-Pères, Paris Cedex, France*

MARK D. TARN • *Department of Chemistry, The University of Hull, Hull, UK*
YUKSEL TEMIZ • *IBM Research GmbH, Rüschlikon, Switzerland*
BRUNO TESTE • *Macromolecules and Microsystems in Biology and Medicine, Institut Curie, Centre National de la Recherche Scientifique, Université Pierre et Marie Curie, PSL Research University, UMR 168, Paris, France*
MANABU TOKESHI • *Department of Applied Chemistry, Graduate School of Engineering, Nagoya University, Nagoya, Japan; Division of Biotechnology and Macromolecular Chemistry, Hokkaido University, Sapporo, Japan*
LISA TRIPODI • *MeBioS - Biosensors, Department of Biosystems, KU Leuven, Leuven, Belgium*
JEAN-LOUIS VIOVY • *Macromolecules and Microsystems in Biology and Medicine, Institut Curie, Centre National de la Recherche Scientifique, Université Pierre et Marie Curie, PSL Research University, UMR 168, Paris, France*
SERGEY ZELENIN • *Science for Life Laboratory, Division of Proteomics and Nanobiotechnology, KTH Royal Institute of Technology, Stockholm, Sweden*

Part I

Chapter 1

Microfluidics-Enabled Diagnostic Systems: Markets, Challenges, and Examples

Holger Becker and Claudia Gärtner

Abstract

Microfluidics has become an important tool for the commercial product development in diagnostics. This article will focus on current technical demands during the development process such as material and integration challenges. Furthermore, we present data on the diagnostics market as well as examples of microfluidics-enabled systems currently under commercial development or already on the market.

Key words Microfluidics, Commercialization, Product development, Diagnostic cartridge

1 Introduction

Microfluidics has made a tremendous progress in the last 25 years since the formulation of the μ-TAS concept by Andreas Manz and coworkers [1]. The scientific merit of miniaturization became apparent very quickly in the different scientific communities in chemistry and the life sciences as apparent by the ever-increasing number of publications in the field (Fig. 1) [2]. Translating these research results into commercial products, namely in diagnostics, has however proven to be a challenging task due to a variety of reasons:

1. Microfluidics itself is an "enabling technology" such as microelectronics and not a product in itself. Any scientific work on the basics of the technology therefore first has to be translated to a suitable application case, which can be a time-consuming step and requires translating scientific questions into an engineering solution [3]. Especially, the applicability of many basic microfluidic concepts (mixing, droplet generation, separations, etc. to name just a few) to a broad range of application fields makes this step often difficult.
2. For practical product development, each microfluidics-enabled system has to be developed largely from scratch. There are almost no official and only a few semi-official standards in

Valérie Taly et al. (eds.), *Microchip Diagnostics: Methods and Protocols*, Methods in Molecular Biology, vol. 1547, DOI 10.1007/978-1-4939-6734-6_1, © Springer Science+Business Media LLC 2017

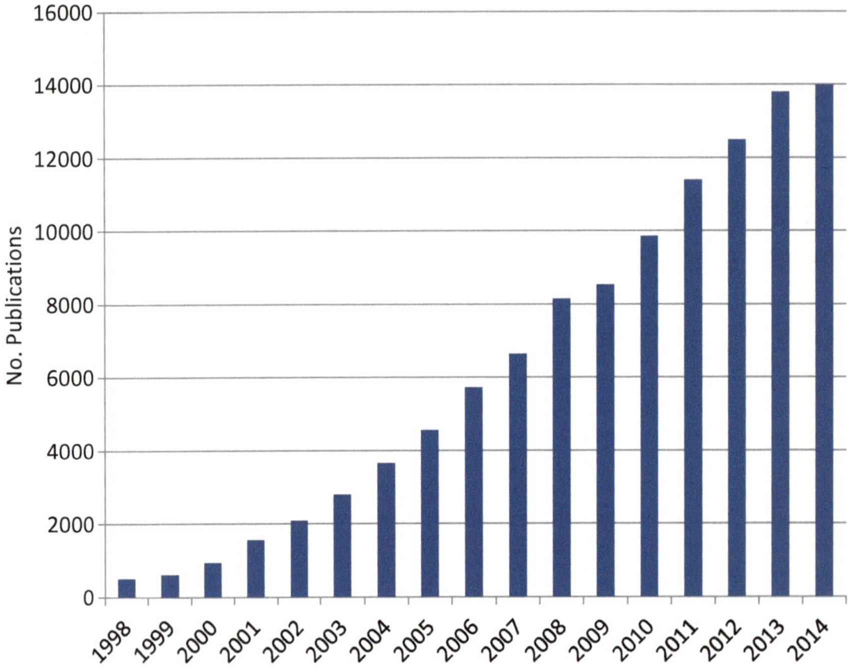

Fig. 1 Development of scientific publication numbers in microfluidics. Data generated by search on Google Scholar with search term "microfluidics" in full text. Data accessed December 10, 2015

microfluidics [4–8], which poses a challenge during the development process. There are efforts to create certain functional modules or platforms [9, 10] (*see* also below in Subheading 2.2.3) which can function similar to a component library in microelectronics to speed up product development; however, these approach have not reached a level of maturity that can be found in other industrial areas such as in labware with its multitude of standardized wellplates or tubing, or in medical technology with threads and interconnects.

3. The lack of these basic tools previously created a significant entry barrier for the use of microfluidics in life science/diagnostic applications. Due to the absence of commercially available off-the-shelf components, for each simple experiment or just a feasibility test prior to the decision to start a product development, a rather extensive engineering project had to be started, involving the design of the respective microfluidic component, the custom manufacturing, and subsequent functional validation. Given the high probability that such an initial design would not be working to the complete specification required (first designs still usually do not), this resulted in a very high hurdle in using microfluidics for practical product development.

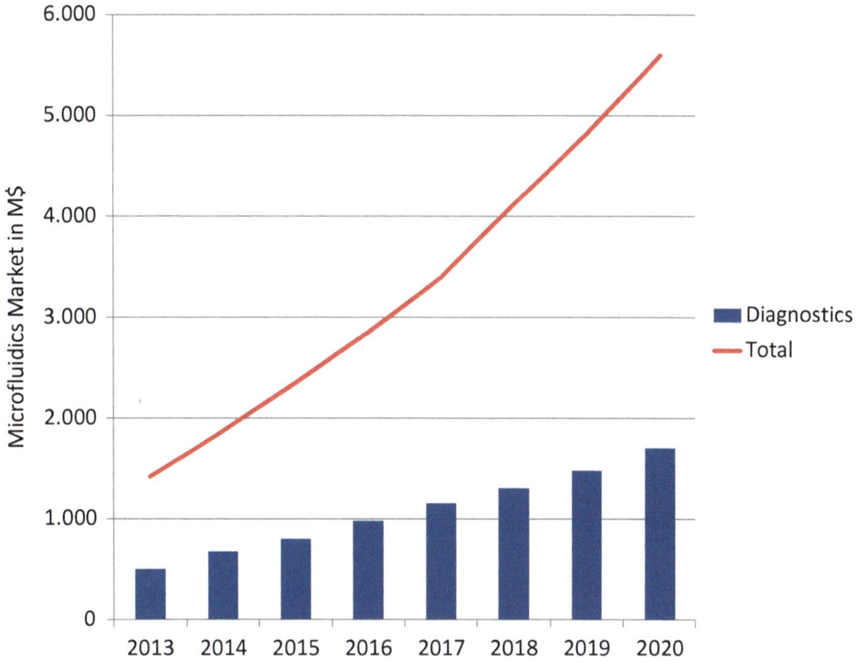

Fig. 2 Microfluidics market volume and segments (adapted from [13])

Despite these difficulties in moving microfluidics technologies from the scientific workspace into the realm of commercial products, the advantages of microfluidics are so compelling that by now a significant number of microfluidics-enabled products have entered the market or are about to do so. In this context, it should be noted that this article concentrates on devices in which the sample is transported through a microstructure-engineered substrate (a.k.a. "cartridge"). Lateral flow devices, where the sample transport happens by capillary forces in a "simple" piece of paper (e.g., nitrocellulose) [11] such as common pregnancy tests or glucose test strips [12], are not subject of this review. The importance of microfluidics-enabled diagnostic devices is reflected in available market research data, *see* Fig. 2 [13]. Also obvious from these market analysis figures is the importance of the diagnostics market as one of the key markets for microfluidics, with about one third of the overall market being represented in point-of-care (POC) or clinical diagnostics. The challenge in generating such market data can be seen in Fig. 3, where the market data from two succeeding market studies [13, 14] have been plotted. It is a typical phenomenon in emerging high-tech markets that market predictions are too optimistic [15], as there is a lack of good predictive tools in industrial practice as well as the psychological factors involved. Nevertheless, it becomes obvious that the diagnostic market is one of the key drivers in the commercialization of microfluidics [16–19].

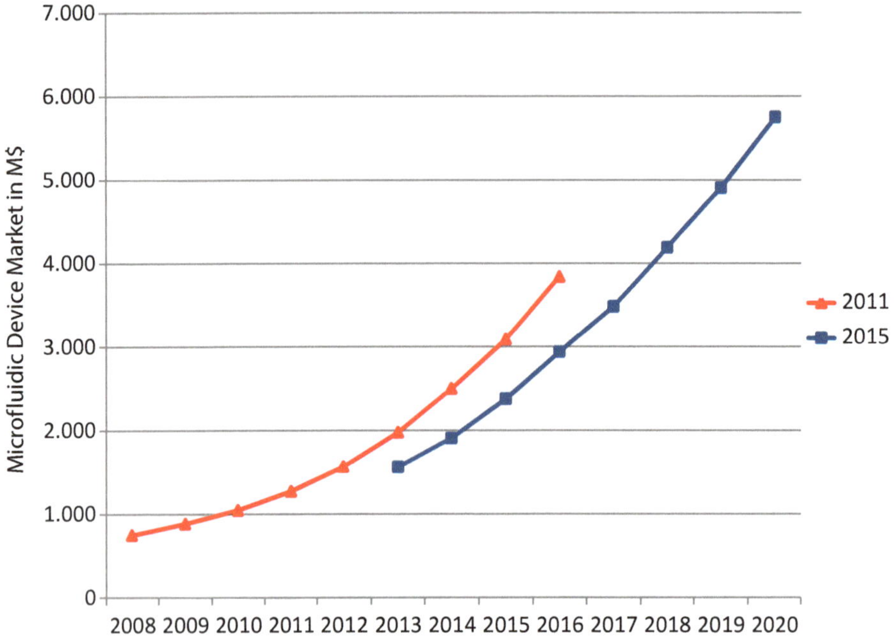

Fig. 3 Comparison of market data from subsequent market studies (adapted from [13] and [14])

To understand the current status of the role of microfluidics in the diagnostic product commercialization process, it is important to analyze the factors behind this development. There are several key elements, which indicate that a crucial stage in the maturing of microfluidics technology had been reached in recent years. This point in time can be characterized by the following properties:

1. Compared to products from the early stages of microfluidics commercialization such Caliper's LabChip [20], the complexity of the devices, especially for diagnostic applications, has dramatically increased. It is the ability to fulfill one of the very early promises of the initial μ-TAS concept, the complete integration of a complex workflow, which had a dramatic influence of the industrial usability of microfluidics [21]. While on the academic side many activities have centered on the development of very specific chip-based protocols, the typical experiment often resembles more a "chip-in-a-lab" situation with a microfluidic device being surrounded by a complex periphery consisting of a pumps, valves, flow controllers, etc. From an industrial point-of-view, a commercially viable product has to provide sample-in answer-out capabilities to compete with existing diagnostic solutions. Especially, sample-preparation steps for nucleic-acid based tests (NAT) form an indispensable part of the diagnostic protocol which has to be performed on-chip.

This has for many years been a somewhat neglected field and only in recent years received more attention [22–26], finally leading to a knowledge-base that allows for true "lab-on-a-chip" solutions necessary for commercial use. Subheading 2 will highlight some key elements necessary to perform the development of such highly integrated microfluidics-based product development.

2. The above-mentioned knowledge-base or a critical mass of know-how has also been generated over the years with respect to manufacturing method and materials. In the early years of microfluidics, most of the devices were made using methods and instrumentation related to the microelectronics industry (lithography, etching of silicon, or glass-based substrates) that, not only due to their cost structure but also due to their geometrical limitations, would have to overcome significant hurdles for any commercial use. It was the increasing use of thermoplastic polymers, notably high-performance polymers such as cyclo-olefin polymers (COC, COP, tradenames "Topas," or "Zeonor") and the associated manufacturing technologies such as hot embossing and injection molding which paved the way for low-cost high-volume manufacturing [27–29]. Besides the economic considerations concerning manufacturing cost, it is important to notice that for any commercial production technology the ability to cover a large dynamic range of structural dimensions is of great importance. While lithography-based technologies allow the generation of very small structures with critical feature sizes and heights in the range of usually micrometers to several tens of micrometers, these technologies are limited to generating one height per lithography step with a maximum height given by the limits of the subsequent etching process (often in the range of several tens of micrometers). For real-life diagnostic protocols, however, which often require comparatively large sample or reagent volumes, it is important to have manufacturing technologies at hand which can create small structures as well and often in direct vicinity of very large (millimeter-sized) structures. In addition, free-form structures such as ramps and arbitrary wall-angles should be possible. Figure 4 shows some examples of such devices with representative geometrical features. Only by emancipating itself from the manufacturing technologies associated with microelectronics and adopting high-volume polymer fabrication technologies such as injection molding, microfluidics was able to enter the mainstream product development path in diagnostics.

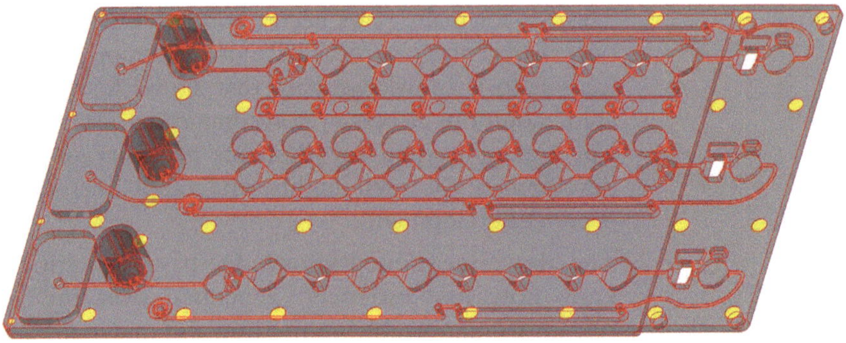

Fig. 4 Example of a microfluidic device with a large dynamic range of geometrical features. Structural elements as small as 100 μm are made in the same step as chambers with 2 mm depth or fluidic connectors of more than 5 mm in height

2 Technical Challenges in the Product Development

2.1 The Complexity Challenge

A fully integrated cartridge has to be able to contain all protocol steps of the diagnostic process. This simple statement, however, poses the biggest challenge on the engineering side of the product development process. If one considers, e.g., a molecular diagnostic protocol, typical process steps are:

- Sample uptake and preconditioning steps, such as filtration, homogenization, or metering.
- Sample preparation steps such as lysis and nucleic acid extraction.
- Amplification of the nucleic acids, e.g., by conventional PCR, isothermal amplification, or other means.
- Separation step, e.g., by electrophoresis, affinity capture, or forces such as magnetic or centrifugal.
- Detection of the analyte of interest, using, e.g., optical (fluorescence, colorimetric, spectroscopic) or electrochemical detection methods.
- Waste handling and storage on-chip.

Immunological assays can have a similar complexity, often without an amplification step but with additional washing steps.

While for all the above-mentioned steps in conventional protocols pipetting or other multiple handling steps (either manual or robotic) are carried out sequentially and typically without interaction, successfully transferring these steps onto a microfluidic cartridge requires a specific approach. The classical engineering approach to a complex problem, the division of the complex problem into individual "smaller" problems, solving of each of these individual elements and the simple superposition of the individual solutions to a system solution, which works, e.g., extremely well in microelectronics, does not work in microfluidic systems. The main reason for this is the interaction of the different elements found in the microfluidic cartridge [11, 21, 30], where each functional element influences the performance of the subsequent elements. Furthermore, the available simulation and modeling tools in microfluidics despite recent advances have limitations in complete system simulation. While being able to model individual elements such as mixers or capillary stop valves [31, 32] or phenomena like heat transfer with high precision [33], the current tool find their limits in system modeling [34]. Therefore, a strategy for the development of such integrated microfluidic devices is required which takes into account these constraints [35, 36]. A two-prong approach has proven to be advisable. On the one hand side, a holistic top-down approach from the system level is necessary to ensure the inclusion of all necessary functions as well as the definition of all interfaces (fluidic, mechanic, optic, etc.) between the microfluidic cartridge and the corresponding instrument. A flow-diagram of all process steps performed on the device can then be translated into individual functional modules. The second line of approach is then a development (e.g., by simulation and subsequent prototyping) of the individual module (e.g., a DNA extraction chamber, a mixing structure for the lysis buffer, etc.) where the individual functions can be validated before integration. It has proven beneficial as best practice in the development to combine the theoretical (or modeling) approach with some experimental data from module prototypes. Once these functions have been verified in the different modules (for examples, *see* Fig. 5), a stepwise integration into

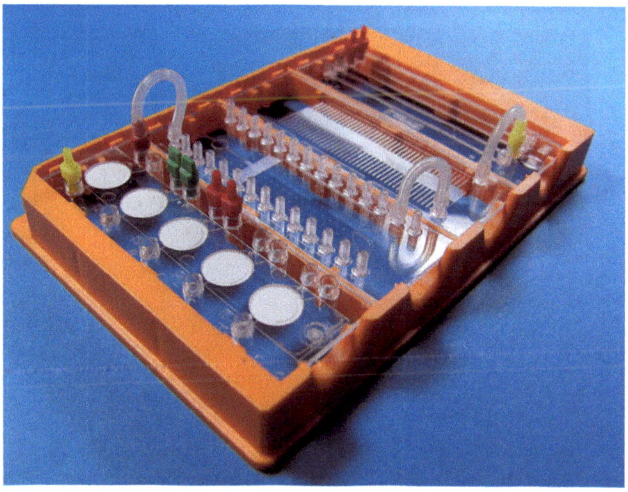

Fig. 5 Examples of microfluidic modules for unit operations such as filtration, separation, amplification, and incubation

a single device then can take place. This stepwise approach also simplifies the search for and correction of possible errors observed in the performance of the device, as troubleshooting of a complex integrated microfluidic circuit is extremely difficult.

2.2 Specific Technological Solutions and Trends

While developing a microfluidic device for diagnostic applications, besides the more principle question of material choice [37, 38], a range of specific technological challenges arises which have to be addressed. These challenges also indicate, in which directions the technical solutions are developing. These technical solutions in turn give indications for application- or user-induced trends, namely with regards to user and performance expectations.

2.2.1 Reagent Storage

In most cartridges, some or all of the reagents used for the diagnostic protocol will have to be stored on-board [39]. The reason for this is an easier handling, less or no fluidic interfaces between cartridge and instrument, leading to a simpler (and most likely smaller) instrument and a reduced chance of contamination. Reagents can be stored either in dry or liquid form. Depending on the volumes needed, liquid reagents, mostly buffers of different kinds, can be stored in blisters, stickpacks, or reagent packs (also called "tanks"). Blisters have been used to store reagents already in one of the first microfluidic products, the i-Stat cartridge [40]. Typical volumes range from 25 μL to about 500 μL. Blisters can be made out of a simple polymer pouch [40, 41] or a thermoformed polymer compound film (Fig. 6) [42–45]. Two important considerations on the system design are the principle of blister opening and the actuation mechanism of the blister. For blister opening, two technologies are available, frangible seals and piercing needles.

Fig. 6 Example of a chip with different thermoformed blisters (volumes between 75 and 300 μL) for liquid reagent storage

In case of a frangible seal, the sealing area between top and bottom film of the blister is structured in a way that at a given pressure, the blister bursts at this predefined location. The challenge in this approach is the risk of having a nonuniform flow rate, with a large spike directly after rupturing of the seal [46]. This has to be compensated by a flow restriction element in the cartridge design which however requires real estate on the cartridge. Therefore, the more frequently chosen method for blister rupturing is the use of needles (Fig. 7a) onto which the blisters are lowered and which pierce the bottom film at a defined force (Fig. 7b). This leads to a more constant flow rate out of the blister. For blister actuation, three main mechanisms exist. The simplest one but also most inaccurate actuation is done by simply manually pressing on the blister with a finger. Obviously, the flow rate in such a case is very variable with the individual force applied by the user. A more accurate method is the inclusion of a clamshell mechanism, either on the cartridge or

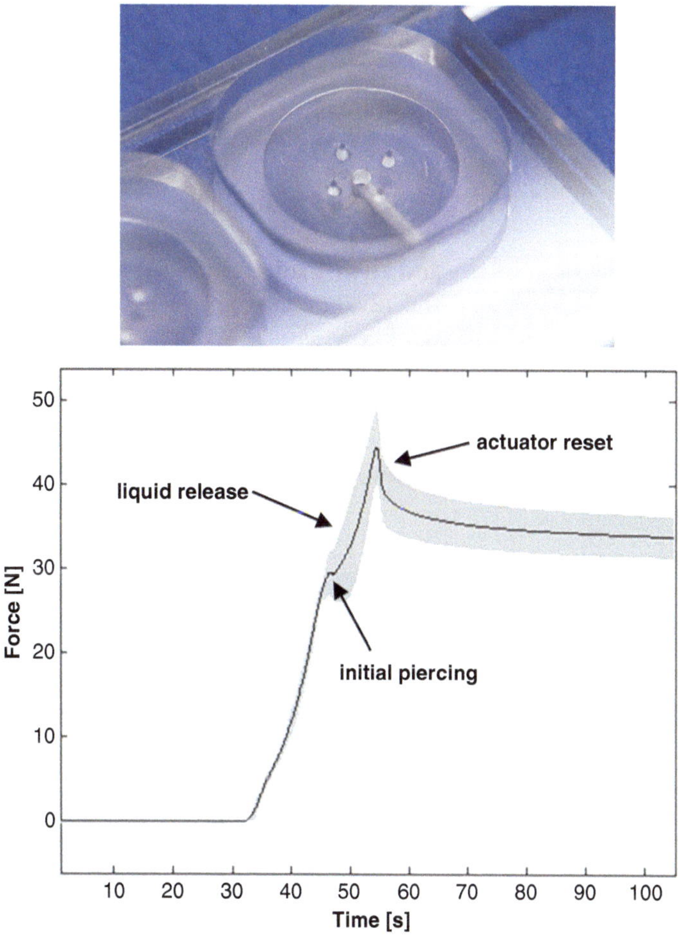

Fig. 7 (**a**) Microneedles embedded in a microfluidic device for blister piercing (**b**) Force diagram for the piercing of a 200 μL blister on the structure shown in (**a**)

the instrument. In this case, the closing of the clamshell generates the pressure to open the blister; however, the mechanism homogenizes the pressure applied manually, leading to a more homogeneous outflow. The most accurate way of emptying a blister is a mechanical actuator in the instrument. This however adds cost, complexity, and volume to the instrument, so this has to be carefully considered during system specification.

Another method for liquid reagent storage on-chip, especially in conjunction with a centrifugal microfluidics platform, is the use of stick packs [47]. These are sachet-like containers with a frangible seal, which burst at a given rotational speed. The possibilities given by the rotating platform in this case mitigate some of the problems discussed with the use of frangible seals.

A third method, especially in case of larger volumes or for sample extraction out of objects like buccal swabs, is the use of reagent packs or "tanks" as shown in Fig. 8. In this case, the reagents are in separate containers, which can be "clicked" onto the cartridge. These containers can contain in addition to the liquid reagents other functional elements such as magnetic beads or filters. The liquid is moved onto the cartridge either by applying air pressure in the tank, a mechanical plunger or simple gravitational flow. Alternatively, negative pressure through the cartridge can be applied.

The storage of dried reagents can be especially useful for molecular diagnostic tests, e.g., in case of PCR master mix reagents [41, 48]. These reagents, often combined with a stabilization agent, can be lyophilized either on-chip [49] or off-chip and then loaded onto the cartridge. On-chip lyophilization has the advantage of a reduced number of steps in the manufacturing workflow as well as the potential to deposit very small reagent volumes. However, the manufacturing throughput is reduced (a cartridge needs significantly more space in a lyophilizer than just a well in a multiwall plate) and for multiplexed diagnostics it might be impossible to find lyophilization conditions, which

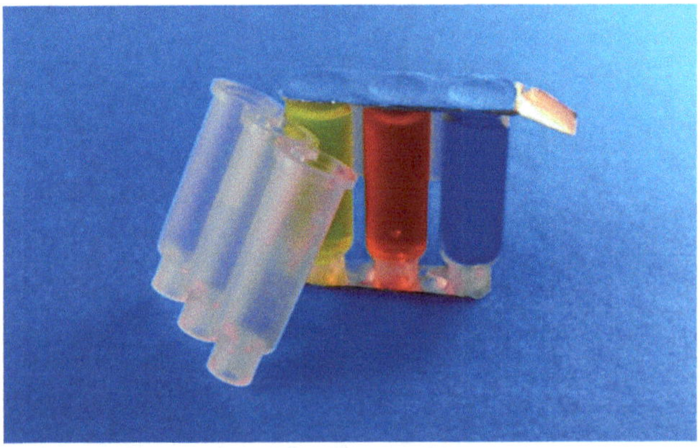

Fig. 8 Reagent tanks with 500 μL volume each for liquid reagent storage

work, for all reagents. Therefore, more frequently, the reagents are lyophilized off-chip in a wellplate and then deposited, e.g., with a pick-and-place robot into the microfluidic cartridge.

All these methods give rise to a clear trend toward a complete reagent integration in the diagnostic cartridge. This allows a system design without any liquid interface between the instrument and the disposable cartridge. The advantages of such a solution are a simplification of the instrument and the reduced risk of contamination as well as an increase in operational safety, as the risk of leakages or imperfect seals at these interfaces is eradicated.

2.2.2 Sensors and Hybrid Integration

Historically and in many cases still today, optical detection technologies define the gold standard for the detection of diagnostically relevant analytes [50, 51]. The reasons for this can be found in the high sensitivity of optical method such as fluorescence as well as the broad range of established assays using these technologies. Limitations arise from the need of a label, which can increase the complexity of the assay and the possible complexity of the optical system, at least in the case of discrete optical components. Electrochemical (for a recent review, *see*, e.g., [52]) or electronic (CMOS [53, 54], GMR [55, 56], etc.) sensors have been investigated as an alternative detection approach. From the industrial manufacturing perspective, this introduces the need for fabrication methods, which can combine heterogeneous materials, e.g., an injection-molded cartridge with the microfluidic structures and a sensor component, which can be a screen-printed electrochemical sensor on another polymer substrate such as PVC, a silicon chip mounted on a printed circuit-board (PCB) or thin-film electrodes on a glass substrate. A similar situation can be found in devices which are used for the spectroscopic analysis of samples, e.g. using Raman spectroscopy [57]. In this case, a polymeric microfluidic cartridge usually gives a Raman background signal that is high compared to the Raman signal coming from the analyte. Therefore, an optical window, ideally made out of materials such as quartz or CaF, has to be integrated into such a cartridge. While for academic settings or prototype production the use of adhesives or lithographically structured photoresists as an intermediate connection layer is a possible solution, in most cases for a high-volume industrial manufacturing, a direct assembly method is preferred to avoid additional materials and handling steps that can deteriorate the production yield [58]. The technologies necessary for these hybrid integration tasks are not well explored in the academic community as they address very specific application cases and are usually only relevant in higher volume manufacturing settings. Such process-related protocols in industry are usually kept as in-house manufacturing know-how and are not published. This generates a challenge in the transition from devices designed in academic groups to a commercial product.

2.2.3 Platform Strategies

From an economic standpoint, it is highly desirable to provide more than a single assay or diagnostic test for a commercial diagnostic system to have a better utilization of the hardware as well as a diversified revenue stream from cartridge consumables. This strategy has been followed by a large number of companies, which have over the years developed complete diagnostic panels for their instrument. A classical example is the iSTAT system [59] from Abbott or, more recently, the GeneExpert [60] system from Cepheid, which started with single assays and expanded the range of applications over the years. The main limitation in these systems lies in the fact that they are limited to a single type of diagnostic assay, i.e., a PCR-based molecular diagnostic assay in case of the GeneExpert or clinical assays in case of the iSTAT system. A truly universal platform, however, should be capable of carrying out different types of assays, i.e., clinical, immunological, molecular diagnostic, and cell-based assays on a single instrument platform. The technical challenges for such an approach are mainly twofold: On the instrument side, different detection technologies for the different assay types have to be incorporated into the instrument design, while on the cartridge side, a common architecture for the cartridges has to be developed. Such an architecture defines main functional areas of a cartridge such as areas for reagent storage, mixing, thermal treatment, or detection as well as all interfaces (mechanical, optical, electrical) between cartridge and instrument. From this architecture, design rules for the microfluidic circuit can then be derived which allow the development of the individual cartridges. Fig. 9 gives an example of a set of cartridges developed

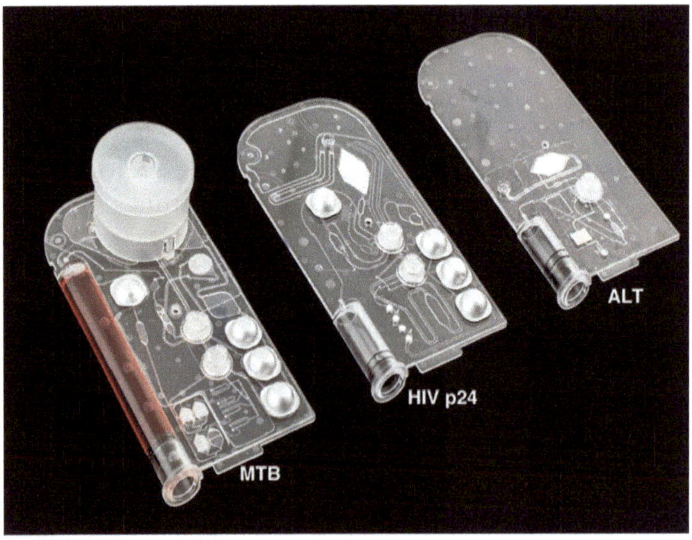

Fig. 9 Set of microfluidic cartridges for a universal diagnostic platform. From *left* to *right*: cartridge for MTB molecular assay, cartridge for HIV p24 immunoassay, and a cartridge for an ALT clinical assay

for such a universal platform [61], from the left to right a cartridge for a molecular diagnostic test for *Mycobacterium tuberculosis* (MTB), a cartridge for an HIV p24 immunoassay and a cartridge for an ALT liver function test as an example of a clinical assay. The common footprint and locations, e.g., for blisters and valves can clearly be seen. Such a system is conceptualized as an open innovation platform, allowing companies to rapidly adopt their specific diagnostic assays onto this platform without the need for instrument modifications. This approach is specifically targeted toward reducing the costs and time required of developers and users to deploy new assays in low-resource settings.

3 Commercial System Examples

In the following subheading, a few examples of microfluidics-enabled system development will be shown to indicate the breadth of current commercial activities, underpinning the market expectations described in Subheading1. A further description of companies using microfluidics-enabled systems in diagnostics can be found in [62]. The first example is the system Evaluation™ from the company MyCartis, which uses a microfluidic cartridge in combination with encoded disc-shaped silicon microparticles (Fig. 10). In its current format, the system is used as an open platform for biomarker research but will be used as a diagnostic platform in future iterations [63]. A second example, in the form of a handheld point-of-care immunoassay system, is shown in Fig. 11. This system from Philips Healthcare, called Minicare, utilizes magnetic nanoparticles [64] to perform a sandwich immunoassay. As the analytes are bound to the magnetic particles, no liquid movement is necessary.

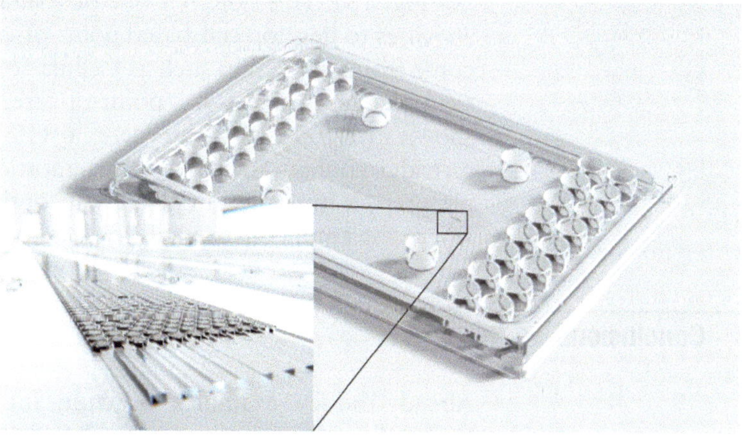

Fig. 10 Microfluidic cartridge with encoded silicon microdiscs for the MyCartis Evaluation™ platform. Image courtesy of N. Demierre, MyCartis

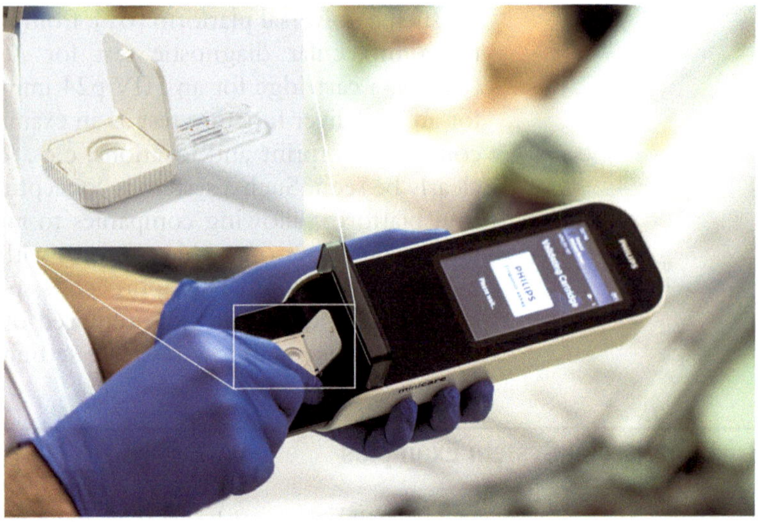

Fig. 11 Point-of-care system Philips Minicare. Image courtesy of J. Nieuwenhuis, Philips Healthcare

The transport is carried out by a moving magnet in the system, dragging the magnetic particles to the desired location. Readout-principle is a variation of total internal reflection (TIRF) [65]. This system is designed as an open platform for an easy third-party assay implementation. Another example for a handheld instrument performing a nanoparticle-based immunoassay is shown in Fig. 12. This system from the UK-based company AgPlus Diagnostics utilizes silver nanoparticles, which are moved onto an electrode array where an electrochemical detection takes place [66]. Again, all required reagents are stored in liquid and dried format on the cartridge. An example for the trend toward cell-based diagnostics is the compact hematology analyzer from the company Ativa Medical (*see* Fig. 13). This system is one of several examples of commercial activities to develop cell-based point-of-care system. Other systems come from companies such as Cellula [67] or PixCell [68]. A final example, in this case for point-of-care molecular diagnostics, is shown in Fig. 14. It is the Alere™ q HIV-1/2 Detect system. This cartridge contains the complete diagnostic workflow including sample prep, a multiplex PCR, and an optical detection of an array, where the amplicons are hybridized.

4 Conclusions

Already the few examples shown in Subheading 3, together with the market data presented in Subheading 1, show the strong commercial interest in microfluidics-enabled diagnostic systems [69]. It can be expected that the trends toward moving the diagnostic

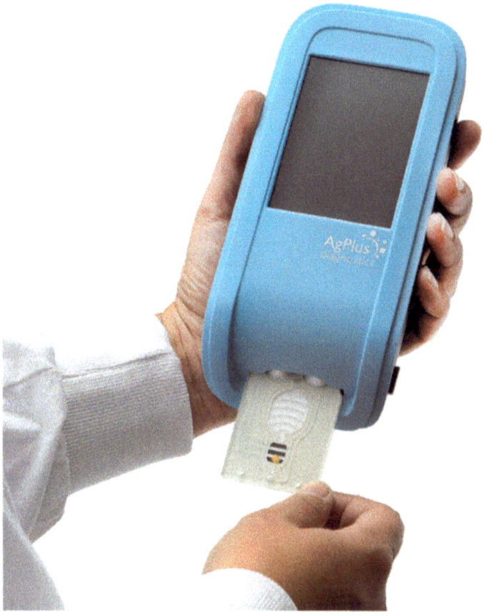

Fig. 12 Handheld immunoassay system from AgPlus Diagnostics. Image courtesy of K. Page, AgPlus Diagnostics

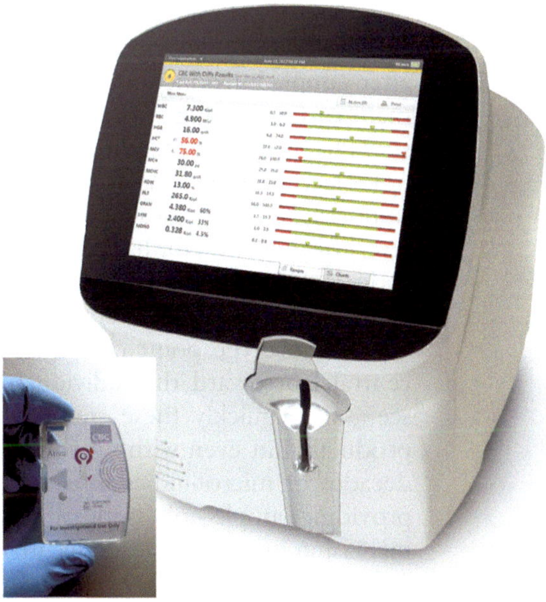

Fig. 13 Example of a cell-based hematology point-of-care system. Image courtesy of S. Floe, Ativa Medical

step closer to the patient as well as the need for a faster and more specific diagnostics in the wake of a raised awareness toward increased antibiotic resistance in many diseases (most prominent in tuberculosis, but also in hospital-acquired infections) will fuel the development of such systems in the years to come. Microfluidics as an enabling-

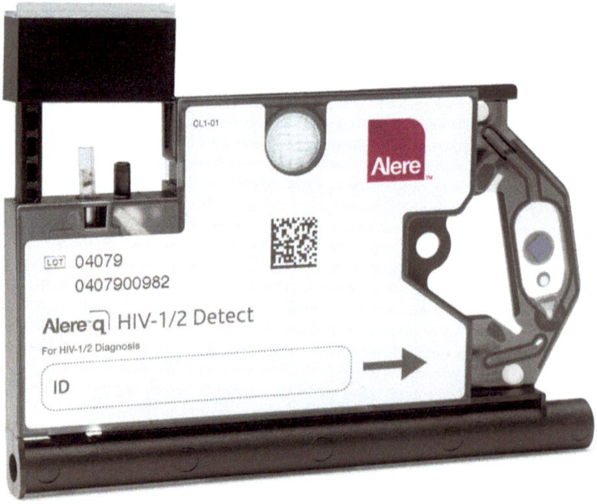

Fig. 14 Alere™ q HIV-1/2 Detect cartridge as an example of a molecular diagnostics point-of-care system. Image courtesy of K. Schindlbeck, Alere Technologies

technology has certainly reached a state of maturity in the sense that also very complex workflows can be integrated in a single cartridge that can be operated without any additional hands-on time after sample introduction. This is a significant step in the development of a technology, when the success of a product development starts to be more depending on market issues rather than on technological challenges. As pointed out in Subheading 2, there remain significant technological challenges to be overcome, also in finding cost-effective ways for high-volume manufacturing of complex cartridges. A further input for future developments comes from concepts associated with wearable and mobile health devices which would move the point of diagnostics even further away from a central lab toward the individual person [70]. It remains to be seen, how quickly these trends will manifest themselves in real products. But even without them, the commercial exploitation of decades of microfluidics research in diagnostics is in full swing, providing ample job opportunities for graduates in the ever-increasing number of academic groups in microfluidics research.

Acknowledgments

Part of the work was carried out within the frame of the EU-FP7 project ROUTINE, contract no. 304941 (www.routinefp7.eu). We thank Nicolas Demierre, Keith Page, Jeroen Nieuwenhuis, Stephen Floe, and Klaus Schindlbeck for providing the images for Subheading3.

References

1. Manz A, Graber N, Widmer HM (1990) Miniaturized total chemical analysis systems: a novel concept for chemical sensing. Sens Actuators B-Chem 1:244–248
2. Yetisen AK, Volpatti LR (2014) Patent protection and licensing in microfluidics. Lab Chip 14:2217–2225
3. Becker H (2010) Lost in translation. Lab Chip 10:813–815
4. Klapperich CM (2009) Microfluidic diagnostics: time for industry standards. Expert Rev Med Devices 6:211–213
5. Becker H (2010) One size fits all? Lab Chip 10:1894–1897
6. van Heeren H (2012) Standards for connecting microfluidic devices? Lab Chip 6: 1022–1025
7. Zhixiong H, Wenli L, Baoyu H et al (2014) Metrological standardizing for future microfluidic-based point-of-care diagnostic products. Sens Transducers 173:250–255
8. van Heeren H, Tantra R, Salomon P (2015) Microfluidic devices: a road forward by standardization of interconnects and classification. Microfluid Nanofluidics 19(5):1203–1207
9. Gärtner C, Becker H, Anton B et al (2004) Microfluidic toolbox: tools and standardization solutions for microfluidic devices for life sciences applications. Proc. SPIE 5345, Microfluidics, BioMEMS, and Medical Microsystems II, pp 159–162. doi:10.1117/12.538373
10. Mark D, Häberle S, Roth G et al (2010) Microfluidic lab-on-a-chip platforms: requirements, characteristics and applications. Chem Soc Rev 39:1153–1182
11. Yetisen AK, Akram MS, Lowe CR (2013) Paper-based microfluidic point-of-care diagnostic devices. Lab Chip 13:2210–2251
12. Clarke SF, Foster JR (2012) A history of blood glucose meters and their role in self-monitoring of diabetes mellitus. Br J Biomed Sci 69(2): 83–93
13. Yole Developpement, Microfluidic Applications 2015. http://www.i-micronews.com/component/hikashop/product/p2015-microfluidic-applications-in-the-pharmaceutical-life-sciences-in-vitro-diagnostic-and-medical-device-markets.html. Accessed 6 Oct 2015
14. Yole Developpement, Emerging Markets for Microfluidic Applications, Report 2011
15. Couillard D (2006) Managing in a sea of uncertainty: leadership, learning, and resources for the high tech firm. Presses internationales Polytechnique, Montreal
16. Yager P, Edwards T, Fu E et al (2006) Microfluidic diagnostic technologies for global public health. Nature 442:412–418
17. Becker H (2009) Hype, hope and hubris: the quest for the killer application in microfluidics. Lab Chip 9:2119–2122
18. Chin CD, Linder V, Sia SK (2012) Commercialization of microfluidic point-of-care diagnostic devices. Lab Chip 12:2118–2134
19. Volpatti LR, Yetisen AK (2014) Commercialization of microfluidic devices. Trends Biotechnol 32(7):347–350
20. Bousse L, Mouradian S, Minalla A et al (2001) Protein sizing on a microchip. Anal Chem 73(6):1207–1212
21. Becker H (2008) Microfluidics: a technology coming of age. Med Device Technol 19(3): 21–24
22. Lagally ET, Scherer JR, Blazej RG et al (2004) Integrated portable genetic analysis microsystem for pathogen/infectious disease detection. Anal Chem 76(11):3162–3170
23. Easley CJ, Karlinsey JM, Bienvenue JM et al (2006) A fully integrated microfluidic genetic analysis system with sample-in-answer-out capability. Proc Natl Acad Sci USA 103(51): 19272–19277
24. Kim J, Johnson M, Hill P et al (2009) Microfluidic sample preparation: cell lysis and nucleic acid purification. Integr Biol 1:574–586
25. Mukherjee S, Kang TG, Chen Y et al (2009) Plasma separation from blood: the 'lab-on-a-chip'approach. Crit Rev Biomed Eng 37:517–529
26. Kersaudy-Kerhoas M, Sollier E (2013) Micro-scale blood plasma separation: from acoustophoresis to egg-beaters. Lab Chip 13:3323–3346
27. Becker H, Gärtner C (2008) Polymer microfabrication technologies for microfluidic systems. Anal Bioanal Chem 390(1):89–111
28. Attia UM, Marson S, Alcock JR (2009) Micro-injection moulding of polymer microfluidic devices. Microfluid Nanofluidics 7(1):1–2
29. Li S, Xu Z, Mazzeo A et al (2008) Review of production of microfluidic devices: material, manufacturing, and metrology. Proc. MEMS, MOEMS, and Micromachining III: Proceedings ofSPIE6993:69930F.doi:10.1117/12.781942
30. Hlawatsch N, Klemm R, Carstens C et al (2012) A lab-on-a-chip system for the development of complex assays using modular microfluidic components. Proc SPIE 8251, Microfluidics, BioMEMS, and Medical Microsystems X, 82510D. doi: 10.1117/12.910269

31. Jen C, Wu C, Lin Y et al (2003) Design and simulation of the micromixer with chaotic advection in twisted microchannels. Lab Chip 3:77–81
32. Hagmeyer B, Zechnall F, Stelzle M (2014) Towards plug and play filling of microfluidic devices by utilizing networks of capillary stop valves. Biomicrofluidics 198(5). Article 056501
33. Hu G, Li D (2007) Multiscale phenomena in microfluidics and nanofluidics. Chem Eng Sci 62:3443–3454
34. Erickson D (2005) Towards numerical prototyping of labs-on-chip: modeling for integrated microfluidic devices. Microfluid Nanofluidics 1:301–318
35. Marques M, Fernandes P (2011) Microfluidic devices: useful tools for bioprocess intensification. Molecules 16:8368–8401
36. Becker H, Gärtner C (2012) Polymeric microfluidic devices for high performance optical imaging and detection methods. In: Fritzsche W, Popp J (eds) Optical nano- and microsystems for bioanalytics. Springer, Berlin, pp. 271–288
37. Ren K, Zhou J, Wu H (2013) Materials for microfluidic chip fabrication. Acc Chem Res 46(11):2396–2406
38. Zhang X, Haswell SJ (2006) Materials matter in microfluidic devices. MRS Bull 31(2):95–99
39. Hitzbleck M, Delamarche E (2013) Reagents in microfluidics: an "in" and "out"challenge. Chem Soc Rev 42:8494–8516
40. Lauks IR, Wieck HJ, Zelin MP et al (1988) Disposable sensing device for real time fluid analysis. US 5096669 A, priority date 15 Sep 1988
41. Chen D, Mauk M, Qiu X et al (2010) An integrated, self-contained microfluidic cassette for isolation, amplification, and detection of nucleic acids. Biomed Microdevices 12(4): 705–719
42. Disch A, Mueller C, Reinecke H (2007) Low cost production of disposable microfluidics by blister packaging technology. Conf Proc IEEE Eng Med Biol Soc 2007:6323–6326
43. Hugo S, Land K, Becker H (2013) Characterization of microfluidic components for low-cost point-of-care devices. Proc. MicroTAS 2013, Freiburg, pp 461–463
44. Jokerst JV, Floriano PN, Christodoulides N et al (2008) Integration of semiconductor quantum dots into nano-bio-chip systems for enumeration of CD4+ T cell counts at the point-of-need. Lab Chip 8:2079–2090
45. Selvakumar S, Linares R, Oppenheimer A et al (2012) Variation analysis of flow rate delivered using a blister pump. Proc. SPIE 8251, Microfluidics, BioMEMS, and Medical Microsystems X, 82510I. doi: 10.1117/12.907502
46. Inamdar T, Anthony BW (2013) Characterizing fluidic seals for on-board reagent delivery. Proc. SPIE 8615, Microfluidics, BioMEMS, and Medical Microsystems XI, 861516. doi:10.1117/12.2006257
47. van Oordt T, Barb Y, Smetana J et al (2013) Miniature stick-packaging—an industrial technology for pre-storage and release of reagents in lab-on-a-chip systems. Lab Chip 13(15):2888–2892
48. Garcia E, Kirkham JR, Hatch AV et al (2004) Controlled microfluidic reconstitution of functional protein from an anhydrous storage depot. Lab Chip 4(1):78–82
49. Seetharam R, Wada Y, Ramachandran S et al (2006) Long-term storage of bionanodevices by freezing and lyophilization. Lab Chip 6:1239–1242
50. Fritzsche W, Popp J (2012) Bioanalytics, in optical nano- and microsystems for bioanalytics, Springer, Berlin, pp 271–288
51. Pires NMM, Dong T, Hanke U et al (2014) Recent developments in optical detection technologies in lab-on-a-chipdevices for biosensing applications. Sensors 14:15458–15479
52. Rackus DG, Shamsi MH, Wheeler AR (2015) Electrochemistry, biosensors and microfluidics: convergence of fields. Chem Soc Rev 44:5320–5340
53. Lee H, Liu Y, Ham D et al (2007) Integrated cell manipulation system—CMOS/microfluidic hybrid. Lab Chip 7(3):331–337
54. Huang Y, Mason AJ (2013) Lab-on-CMOS integration of microfluidics and electrochemical sensors. Lab Chip 13(19):3929–3934
55. Pekas N, Porter MD, Tondra M et al (2004) Giant magnetoresistance monitoring of magnetic picodroplets in an integrated microfluidic system. Appl Phys Lett 85(20):4783–4785
56. Becker H, Carstens C, Kuhlmeier D et al (2012) Stationary fluidics: moving target molecules on beads through non-moving liquids for molecular diagnostic assays. Proc MicroTAS 2012, Okinawa, pp 791–793
57. Ashok PC, Dholakia K (2012) Microfluidic Raman spectroscopy for biochemical sensing and analysis. In: Fritzsche W, Popp J (eds) Bioanalytics, in optical nano- and microsystems for bioanalytics. Springer, Berlin, pp. 247–268
58. Becker H (2009) It's the economy. Lab Chip 9:2759–2762

59. https://www.abbottpointofcare.com/products-services/istat-test-cartridges. Accessed 8 Dec 2015
60. http://www.cepheid.com/us/cepheid-solutions/clinical-ivd-tests/healthcare-associated-infections. Accessed 8 Dec 2015
61. Becker H, Klemm R, Dietze W et al (2015) Sample-in answer-out point-of-care cartridge for fast MTB diagnostics as part of a universal diagnostics system for global health applications. Proc. MicroTAS 2015, Gyeongju, pp 963–965
62. Kim L (2013) Overview on the microfluidic diagnostics commercial landscape. In: Jenkins G, Mansfield CD (eds) Microfluidic diagnostics, methods in molecular biology, Springer, New York, vol 99. pp 65–84.
63. Laroy W, Ladestein P (2015) MyCartis NV company profile. Biomark Med 9(2):85–88
64. Hefti M, Raymond F, lmmink A et al (2013) Next generation, fast and accurate point-of-care test for NT-proBNP based on Magnotech technology. Point Care 12(4):171
65. Dittmer WU, Evers TH, Hardemann WM et al (2010) Rapid, high sensitivity, point-of-care test for cardiac troponin based on opto-magnetic biosensor. Clin Chim Acta 411:868–873
66. Wilson PK, Szymansk M, Porter R (2013) Standardisation of metalloimmunoassay protocols for assessment of silver nanoparticle antibody conjugates. J Immunol Methods 387:303–307
67. http://www.celula-inc.com/. Accessed 5 Dec 2015
68. http://www.pixcell-medical.com/. Accessed 5 Dec 2015
69. Gubala V, Harris LF, Ricco A et al (2012) Point of care diagnostics: status and future. Anal Chem 84:487–515
70. https://www.aacc.org/publications/cln/articles/2013/may/mobile. Accessed 8 Dec 2015

Part II

Microchips for Protein Bioassays

Chapter 2

Capillary-Driven Microfluidic Chips for Miniaturized Immunoassays: Efficient Fabrication and Sealing of Chips Using a "Chip-Olate" Process

Yuksel Temiz and Emmanuel Delamarche

Abstract

The fabrication of silicon-based microfluidic chips is invaluable in supporting the development of many microfluidic concepts for research in the life sciences and in vitro diagnostic applications such as the realization of miniaturized immunoassays using capillary-driven chips. While being extremely abundant, the literature covering microfluidic chip fabrication and assay development might not have addressed properly the challenge of fabricating microfluidic chips on a wafer level or the need for dicing wafers to release chips that need then to be further processed, cleaned, rinsed, and dried one by one. Here, we describe the "chip-olate" process wherein microfluidic structures are formed on a silicon wafer, followed by partial dicing, cleaning, and drying steps. Then, integration of reagents (if any) can be done, followed by lamination of a sealing cover. Breaking by hand the partially diced wafer yields individual chips ready for use.

Key words Microfluidic chip fabrication, Capillary-driven flow, Dry-film resist, Chip singulation, Sealing

1 Introduction

Microfluidics refers both to a technology and set of devices capable of manipulating and analyzing microliter and submicroliter volumes of liquids and samples [1, 2]. The concept of microfluidics became quickly popular in the mid-1990s, when Whitesides' group at Harvard showed how to mold poly(dimethylsiloxane) (PDMS) layers using lithographically prepared masters. Such PDMS layers could be microstructured and sealed with surfaces to produce microfluidic networks for patterning proteins and miniaturizing immunoassays on surfaces [3]. Since then, many concepts for microfluidic-based assays have emerged for the detection of antigens, nucleic acids, pollutants, and pathogens in body fluids and environmental liquids [4–6] with a strong emphasis on portability [7, 8]. The concept of point-of-care analysis using microfluidic devices is supported by the possibility of integrating complex

Valérie Taly et al. (eds.), *Microchip Diagnostics: Methods and Protocols*, Methods in Molecular Biology, vol. 1547,
DOI 10.1007/978-1-4939-6734-6_2, © Springer Science+Business Media LLC 2017

functionalities (e.g., micropumps, microstirrers, valves, electronic circuitry for reading signals, etc.) on one device, to produce small factor devices and utilize advanced packaging techniques, and to implement faster and more precise tests, which can also be conservative of reagents and samples [9–12].

While the design, fabrication, materials chosen, and protocols for devising microfluidic-based immunoassays can strongly vary and depend on a specific application, some generalities are conserved across applications; (1) microfluidic structures must be created in a layer, (2) these microfluidic structures must be sealed with another layer, (3) reagents and receptors for analytes must be integrated in the microfluidic chip, (4) liquids must be either actively pumped or passively drawn through the microfluidic structures, and (5) chips are preferably prepared by batches and separated (by dicing) from each other. We earlier introduced the concept of capillary-driven chips for immunoassays [13] using wettable microfluidic structures to move one or a few microliters of sample through an area containing fluorescently labeled antibodies and surface-immobilized capture antibodies [14]. This assay implementation is similar to what has been done with lateral flow assays for pregnancy testing [15] as well as many tests done near patients [16]. Many materials and design variants for the chips can be used but we suggest that developing tests and even performing selected applications can be done very efficiently using silicon wafers/chips. Here, we specifically demonstrate how to fabricate microfluidic chips using silicon wafers that are sealed with a layer of dry-film resist (DFR), which is a negative-tone photoresist thick film typically used in microfabrication [17–19]. We also show how to partially dice the wafer early during fabrication so that chips can be cleaned, rinsed, dried, further processed on a wafer level rather than one by one [20]. With this approach, single chips are readily separated from each other by breaking them off the wafer by hand alike when breaking a bar of chocolate. We call this overall fabrication process "chip-olate" and describe it in detail in the subsequent Subheadings. Furthermore, this chapter is complementary to the following one in which we show how to pattern antibody receptors on a DFR using microcontact printing (*see* Chapter 3).

2 Materials

1. Photomask: 5 × 5 inch^2 soda lime glass substrate, Cr coating, 530 nm photoresist layer (Nanofilm, USA) (*see* **Note 1**).
2. Microfluidic substrate: 4 inch (100 mm) Si wafer, N/Phos doping, 1–10 ohm.cm resistivity, <100> orientation, 525 ± 25 μm thickness (Si-Mat, Germany) (*see* **Note 2**).

3. Plasma cleaning: GIGAbatch 310 M resist ashing system (PVA TePla, Germany), O_2 flow: 500 sccm, Power: 600 W, start pressure: 0.5 mbar.
4. Hot plate: C-MAG HS7 (IKA®-Werke GmbH & Co. KG, Germany).
5. Structural microfluidic material: PerMX™3020 (DuPont™), 20 μm thick, epoxy-based, negative-tone, chemically amplified dry-film resist (DFR), protected on both sides with polyolefin and Mylar films (*see* **Note 3**). Store in the dark and preferably in a fridge (4–8 °C).
6. Lamination: Albyco Photopro 33 hot roll laminator (Mega Electronics, UK), 1.5 mm thick Al sheet as a support, and cleanroom paper (80–92 g/m^2, VWR International).
7. Scotch Magic tape (3M).
8. Mask aligner: MJB4 manual mask aligner (SÜSS MicroTec AG, Germany) with 13 mW/cm^2 I-line power density.
9. DFR developer: Propylene glycol methyl ether acetate (PGMEA) (Sigma-Aldrich).
10. Isopropyl alcohol (IPA) (VLSI semiconductor grade, >99.9 %, Technic France).
11. Positive-tone photoresist for protection of the wafer surface during dicing: AZ®6612 (MicroChemicals GmbH, Germany).
12. Dicing tape: 75 μm thick SPV 224P surface protection tape (Nitto Europe).
13. Dicer: ESEC 8803 Dicing Saw with 45 μm width FTB R46 45130 rigid wheel blade (Mithsubishi), 30,000 rpm blade rotation speed and 10 mm/s feed rate.
14. Acetone (VLSI semiconductor grade, >99.5 %, Technic France).
15. Microfluidic sealing layer: DF-1050 (Engineered Materials Systems Inc., USA), 50 μm thick, high performance, chemically amplified, I-line sensitive, negative-tone DFR. Store in the dark and preferably in a fridge (4–8 °C).
16. 0.2 mm thick transparent plastic foil.
17. Electronic craft cutter: Silhouette Portrait® (Silhouette America) with Silhouette Studio® software.
18. Wafer box for storage.

3 Methods

Perform all DFR processing steps under yellow light (UV and blue light filtered), at room temperature and 40–50 % relative humidity, and preferably in a clean room or in a laminar flow hood. Handle DFRs with gloves and wear safety glasses until they are fully exposed and cured to avoid any allergic reaction. Dispose the excess DFR and liquid waste to special chemical waste containers according to disposal regulations.

3.1 Patterning Microfluidic Structures Using a DFR

1. Design a microfluidic chip comprising a circular loading pad, a microfluidic channel, a capillary pump having circular pillars and air vents (*see* Fig. 1a). Prepare a photomask by fabricating a glass/Cr mask using direct laser writing (*see* **Note 4**).
2. Take a 4 inch (100 mm), single side polished Si wafer with a <100> orientation and grow 200 nm thick SiO_2 film using thermal oxidation (*see* **Note 5**). Expose the wafer to O_2 plasma for 2 min and bake for 5 min at 150 °C on a hot plate to clean and dehydrate the surface, respectively.
3. Cut PerMX™3020 DFR using a scissor or blade to about 12 cm × 12 cm size (slightly larger than the wafer).

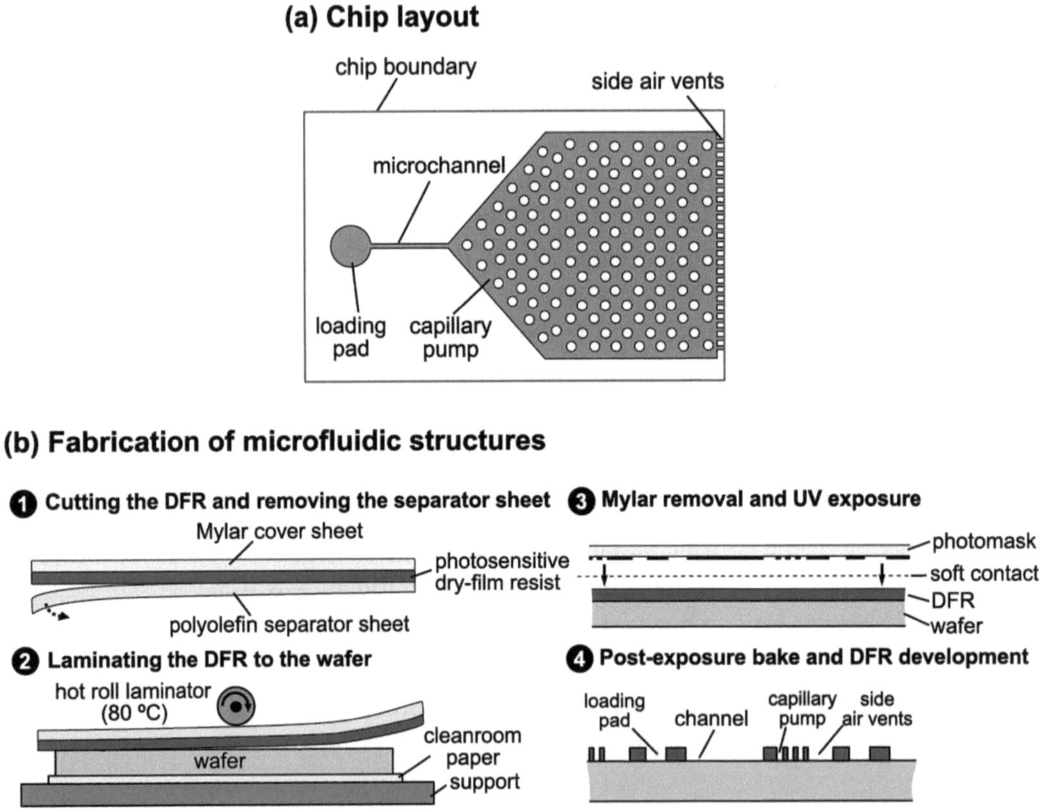

Fig. 1 Main steps describing the fabrication of microfluidic chips before partial dicing. (**a**) A microfluidic chip is designed using an appropriate software. Here, white areas in the design represent areas where the dry-film resist (DFR) will need to be kept to create microfluidic structures. (**b**) A flat substrate such as a Si wafer is laminated with a thin DFR. Local exposure of the DFR to UV prevents dissolution of the exposed areas (negative-tone resist) during a subsequent development wet step. The resulting DFR structures are hardened during a postexposure bake and are 20-μm-thick in the work described here

4. Take an Al sheet (preferably 1.5 mm thick and 20 cm × 20 cm dimensions) and protect the surface of it using a cleanroom paper fixed with a Scotch tape (*see* **Note 6**). Place the wafer on the paper (fixing is not required). Remove the polyolefin protective film on the DFR with the use of a Scotch tape or a tweezer (*see* **Note 7**). Place the DFR on the top of the wafer and fix the upper edge of the DFR to the cleanroom paper using a tape. Feed the metal support to the hot roll laminator set at a temperature of 80 °C and a linear speed of about 30 cm/min. Hold the lower edge of the DFR lifted up and straight, and slowly release while the wafer proceeds in the laminator to avoid formation of air bubbles and wrinkles (Fig. 1b).
5. Following the lamination, cut the excess of the DFR and cleanroom paper along the wafer edges. Keep the wafer on a hotplate at 100 °C for 2 min to enhance film adhesion to wafer surface (post lamination bake). Let the wafer cool for 1–2 min at room temperature.
6. Remove the Mylar film by pulling it from the edge of the wafer using a Scotch tape or a precision tweezer. Expose the wafer to UV light (I-line, 365 nm) through the photomask having the patterns of microfluidic structures using a mask aligner in soft contact mode (*see* **Note 8**). The film requires about 260 mJ/cm^2 exposure dose (e.g., 20 s exposure for a 13 mW/cm^2 light intensity). For Si substrates, the photomask should be aligned as much as possible to the wafer flat so that the dicing cuts be well aligned to the crystallographic planes of the wafer (*see* **Note 9**).
7. Bake the exposed wafer at 100 °C for 2 min on a hot plate (post exposure bake). After 20 s, a visible contrast difference forms between the exposed and unexposed patterns. Let the wafer cool for 1–2 min at room temperature (*see* **Note 10**).
8. Prepare three glass beakers: two of them with about 50 mL PGMEA each and one with IPA (50 mL IPA). Immerse the wafer into the first beaker with PGMEA, develop the film for 60 s while gently shaking the beaker, immerse the wafer into the second beaker with fresh PGMEA and continue the film development for another 30 s, and finally immerse the wafer into the third beaker with IPA. It is important that the liquid should not dry on the wafer surface during the transfer between the beakers (*see* **Note 11**). Rinse the wafer by purging IPA from a squirt bottle and dry using a flow of N_2.
9. Keep the wafer on a hot plate at 150 °C for 10 min (hard bake).
10. Measure the actual channel depth using a surface profiler (*see* **Note 12**).

3.2 Chip-Olate Process: Partial Dicing of Substrate, Cleaning, Lamination, and Singulation of Individual Chips

1. Protect the wafer surface using a thin photoresist layer (e.g., AZ®6612) spin-coated at 4000 rpm for 40 s and baked at 110 °C for 1 min (*see* **Note 13**).
2. Mount the wafer on a dicing chuck using a dicing tape with backside of the wafer facing to the tape. Any adhesion enhancement technique, such as heat treatment, and/or more expensive UV-sensitive dicing tapes, is not required since chip loss during dicing is not probable.
3. Mount the dicing blade to the dicer, calibrate the blade-to-chuck distance, and set the distance to 350 μm, yielding about a 250 μm deep dicing cut for a 525 μm thick Si wafer and 75 μm thick dicing tape (*see* **Note 14**). Dice the wafer along the dicing lines, which are already patterned on the DFR layer (Fig. 2). Rinse the wafer using DI water and dry (N_2 flow, compressed air, or natural evaporation). Cut the tape around the wafer using a scalpel (do not directly remove the wafer from the tape because this action may break the wafer).
4. Place the wafer on a cleanroom wiper with front surface facing to the wiper and slowly remove the dicing tape by gently pulling the tape without bending the wafer. Remove the photoresist protection by immersing the wafer into acetone for 20–30 s and immediately after into IPA without letting the acetone dry on the surface. Remove the wafer from the beaker using a wafer tweezer and purge some IPA from the squirt bottle and dry using a flow of N_2 (Fig. 3).
5. Take DF-1050 DFR, cut it to about 12 cm × 12 cm size, and cut out 2 mm-diameter loading pad openings using an electronic draft cutter such as Silhouette Portrait® (a punch or a cutting die can be also used). Use Silhouette Studio software to define all loading pad openings on the wafer. Place the DFR on a plastic film support (e.g., overhead projector foil)

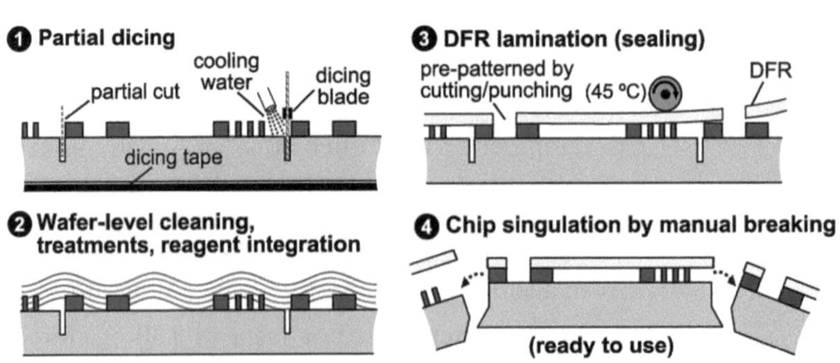

Fig. 2 Key steps illustrating the "chip-olate" concept. Once microfluidic structures are patterned on a wafer, the wafer is partially diced and cleaned. Any surface modification, integration of reagents using an inkjet spotter, and/or lamination of a DFR cover film can be done on a wafer level because single chips will be thereafter separated by breaking the wafer by hand

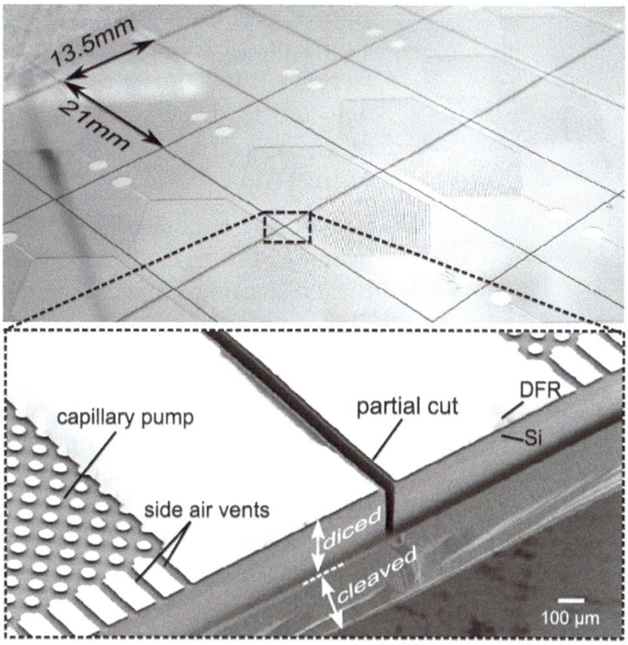

Fig. 3 Optical (*top*) and scanning electron micrograph (*bottom*) showing microfluidic chips after partial dicing. Microfluidic structures are patterned using a 20-µm-deep DFR and a Si wafer

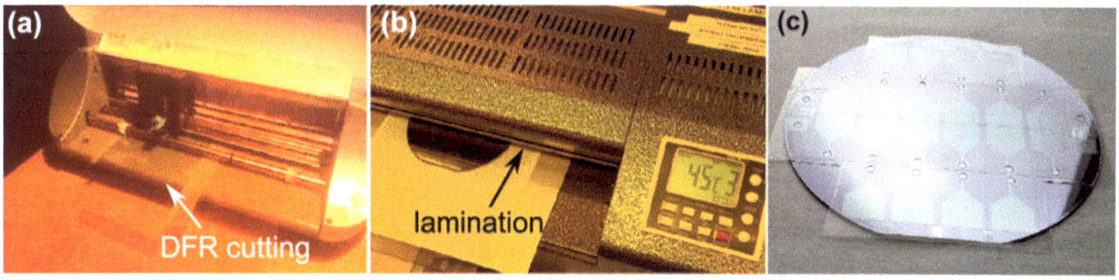

Fig. 4 A DFR film is (**a**) cut using a cutting plotter to define large open areas such as the loading pads and (**b**, **c**) laminated at 45 °C on a partially diced wafer to seal the microfluidic chips

using a Scotch tape and feed it to the cutter (Fig. 4a). Set the cutting speed to lowest and calibrate the blade height using the test cut option. Following the cutting, remove the DFR from the foil.

6. Place the processed, partially diced, and cleaned wafer on the Al support for lamination (*see* **step 4** in Subheading 3.1). Gently blow N_2 or compressed air to remove dust particles on the wafer surface. Remove one of the protective films on the DFR using a Scotch tape or a tweezer and manually align the DFR to the wafer so that loading pad openings on the DFR are

aligned with the loading pads on the wafer (*see* **Note 15**). Place a thick plastic foil (e.g., 200 μm thick) on the top of the DFR to prevent collapsing of the DFR into the microfluidic channels and feed the stack into the laminator at a temperature of 45 °C and a linear speed of 30 cm/min (Fig. 4b, c) (*see* **Note 16**).

7. Expose the laminated wafer to UV light for a total exposure dose of 300 mJ/cm^2 and remove the top protection film on the DFR using a tweezer or Scotch tape (the top protection film can be removed more easily after UV exposure).
8. Bake the DFR by placing the wafer on a hot plate at 75 °C for 30 min (*see* **Note 17**).
9. Singulate the chips by holding the wafer substrate with two hands and gently pressing downward with two thumbs placed two sides of the dicing cut, like breaking a chocolate bar (*see* **Note 18**). Repeat this step until all chips are singulated (Fig. 5). Store the chips at room temperature in a closed wafer box or in heat seal bags (*see* **Note 19**).
10. To test the capillary filling of a liquid, pipette 2–5 μL of liquid to the loading pad (Fig. 6a) (*see* **Note 20**). Liquid is pulled to the channel by a negative capillary pressure, which is calculated to be −4.6 kPa for water at room temperature, for a 200-μm-wide and 20-μm-deep channel and based on measured contact angles of the channel walls. The liquid then reaches the capillary pump at the end of the channel and the pump having a total liquid capacity of 1.9 μL steadily pulls the liquid until it becomes filled (Fig. 6b, c). For this design, the average flow rate is measured as 1.18 μL/min, with a standard deviation of 0.15 μL/min (*see* **Note 21**).

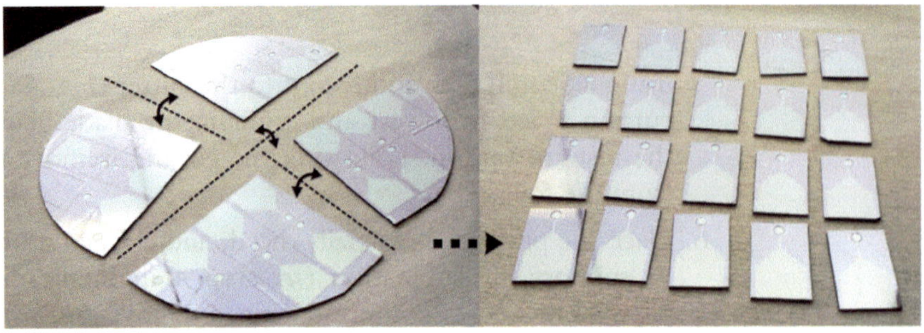

Fig. 5 Illustration of the singulation of individual microfluidic chips by breaking by hand a partially diced wafer patterned with microfluidic structures (photolithographically patterned DFR) and laminated with a layer of DFR (cut using a cutting plotter)

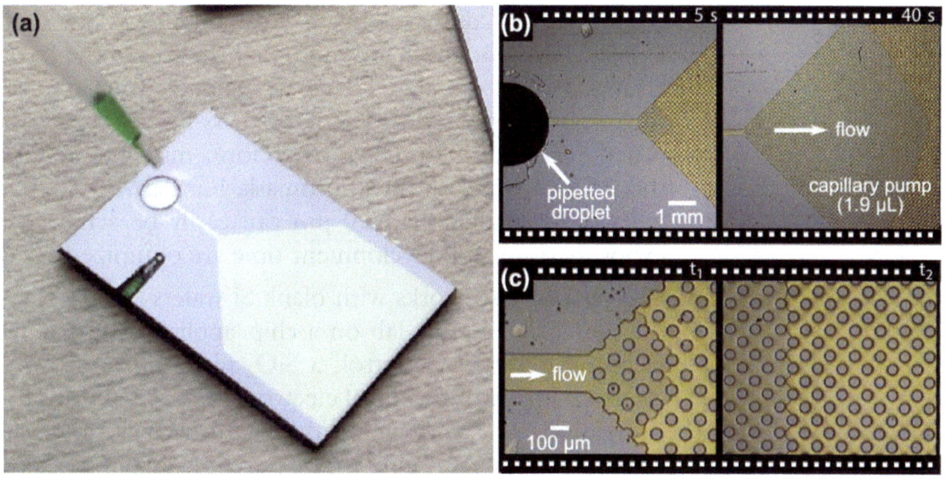

Fig. 6 Functional capillary-filling test of a microfluidic chip using colored water. (**a**) Upon pipetting 2 μL of deionized water, the water spontaneously fills (**b**) a main microchannel inside which tests such as immunoassays may be carried out and (**c**) reaches a capillary pump. The capillary pump represents the largest structure of the chip and can pump about 1.2 μL of liquid over 1 min. Here, small vents at the end of the capillary pump are not visible

4 Notes

1. Wafer layout can be printed on transparent foils using laser printing (limited resolution) or outsourced for a more professional printing on foils. If a direct writing laser system is available (e.g., DWL2000 Heidelberg Instruments, Germany), glass/Cr photomasks can be fabricated with about 1 μm resolution. Microfluidic chips typically have minimum features larger than 50 μm.
2. A glass wafer can also be used as the substrate carrying the microfluidic structures. DFR processing can be applied to any compatible substrate regardless of its size and shape. For Si wafers, doping type and resistivity are not critical for the process.
3. Microfluidic structures can also be patterned using SU-8 (an epoxy-based negative-tone photoresist). Alternatively, they can be fabricated using Si etching (wet etching or deep reactive ion etching) or glass etching in case a glass substrate is preferred. We prefer DFR processing both for fabricating the microfluidic structures and sealing them, because most DFRs are chemically stable, have good adhesion to surfaces (e.g., glass, Si, or cured DFR), allow sufficiently high photolithography resolution, and they require less capital investment compared to other micromachining techniques. Wangler et al. [19] and Ito et al. [21] previously studied cytotoxicity and chemical resistance of similar DFRs and suggested their use in biologically relevant solvents and buffers, and acidic/neutral solutions.

4. Chip layout is drawn using a layout editor (e.g., Cadence, L-Edit, KLayout, etc.) and multiple instances of the layout are arranged to fit a wafer. DFRs are negative-tone photoresists (i.e., regions exposed to UV light cross-links and stays on the wafer during development); therefore, microfluidic channels should be opaque on the photomask. For PerMX™3020, channel widths smaller than 20 μm can easily be achieved if the exposure dose and development time are optimized.
5. The process also works with blank Si wafers (only native SiO_2 on the surface). For lab-on-a-chip applications that require electrical detection/control, a SiO_2 film can be grown as an electrical insulator for metal electrodes patterned prior to DFR lamination. If a thermal oxidation furnace is not available, wafers having a SiO_2 film can be purchased from the supplier.
6. The paper protects the Al sheet from DFR and enables the reusability of it without cleaning. If the sheet is contaminated by DFR, uncured DFR can be cleaned using a cleanroom wiper and acetone.
7. The side with the polyolefin film (translucent) may not be easily detectable by eye but it can be peeled off easier than the Mylar film (transparent) on the other side. Some of the professional laminators do the film removal and feeding automatically.
8. We find that stronger contact modes, such as hard contact or vacuum contact, can result in adhesion of the DFR to the photomask.
9. For Si wafers, rotational misalignments more than a few degrees can result in edge defects during final cleaving for chip singulation. For glass substrates, misalignment is not critical because glass does not have crystallographic planes.
10. Unlike thick SU-8 processing, we find that a temperature profile with slow ramping and cooling is not required for the DFR.
11. A white residue may form on the surface if the wafer is not transferred to fresh PGMEA in the second beaker. Several wafers can be developed using the same beakers as long as the fresh PGMEA is used only for the second development. If the DFR is not fully developed inside narrower channel, this may be due to underdevelopment or overexposure. Film delamination during development might occur in case of underexposure, overdevelopment, and/or insufficient surface cleaning and dehydration. Small cracks can be seen at the sharp edges of the patterns but we find that they disappear during hard baking.
12. For this particular film, we measure the final channel depth as 19.5 ± 0.2 μm using Veeco Dektak 6M surface profiler.
13. Variation in the photoresist thickness due to the topography of the patterned microstructures is not critical for the dicing step.

14. For Si wafers, we find that a dicing cut about the half of the wafer thickness gives the optimal result because wafer is still rigid enough for further processing and a precise breaking can be achieved by applying a small force (e.g., 3 N, measured using a load frame (SELmaxi, Thelkin AG, Switzerland) by placing the samples with the diced face of the substrate resting on two support rods (spaced by 25 mm) and applying the load to the other face). Shallower cuts (<200 μm) require more breaking strength (>4 N) and reduce the efficiency of breaking as some chips may have slightly damaged edges. Wafers with deeper cuts (>400 μm) require more cautious wafer handling for the postdicing steps and the breaking force is not high enough to break both the substrate and the DFR. For glass wafers, a slightly deeper cut, typically 50 μm more than that of Si, is needed to achieve easier breaking because glass does not have crystallographic planes.
15. Wafer-level manual alignment gives sufficient precision (e.g., less than 0.5 mm) for aligning loading pad openings with microfluidic structures.
16. We find that DF-1050 DFR has very suitable mechanical, optical, and wetting properties as a cover layer for microfluidic chips. The DFR shows an excellent tenting (less than 1 μm sagging) for channel widths up to 200 μm and for the capillary pump pillars, and sagging less than 10 μm for much wider channels (wider than 750 μm). It is transparent and optically clear. Our fluorescence spectrophotometer measurements (using Hitachi F-4500 spectrophotometer) show that DF-1050 has autofluorescence intensities as low as those of commonly used PDMS (polydimethylsiloxane) and COC (cyclic olefin copolymer) for wavelengths ranging from 400 to 700 nm. Therefore, microfluidic structures sealed with this DFR can be used in applications based on fluorescence detection of analytes or for monitoring particles/cells in solutions. Advancing contact angle is measured as 63° (using a goniometer from KRUSS GmbH, Germany), suggesting that the film is compatible with capillary-driven microfluidics.
17. Baking temperature and duration should be adjusted according to the application. For example, if the microfluidic chip has integrated bio-reagents, the baking temperature can be lowered or completely skipped. We find that insufficiently UV exposed or insufficiently cured film may slightly deform (several μm) during the liquid flow particularly for channels wider than 100 μm.
18. Wear gloves to avoid any allergic reaction and to contaminate the chip surface. Alternatively, chips can be singulated using two tweezers placed two sides of the dicing cuts and one of them twisted and pressed downward for breaking the substrate.

19. We find that capillary filling works several months after the fabrication when the chips are stored in a wafer box. Do not expose the chips to direct sunlight for longer duration because this may deform the DFR.
20. Loading pad opening patterned on the DF-1050 DFR can hold a few μL of liquid. If more liquid is needed to be pipetted, a plastic piece with a hole can be glued to the loading pad as a reservoir. We find that UV-curable NOA68 (Norland Optical Adhesive) can be used for this purpose.
21. The flow rate can be adjusted by changing the flow resistance of the main channel or the dimensions and arrangement of the pillars in the capillary pump [22].

References

1. Whitesides GM (2006) The origins and the future of microfluidics. Nature 442:368–373
2. Folch A (2012) Introduction to BioMEMS. CRC Press, Boca Raton
3. Delamarche E, Bernard A, Schmid H et al (1997) Patterned delivery of immunoglobulins to surfaces using microfluidic networks. Science 276:779–781
4. Chin CD, Linder V, Sia SK (2012) Commercialization of microfluidic point–of–care diagnostic devices. Lab Chip 12:2118–2134
5. Jokerst JC, Emory JM, Henry CS (2012) Advances in microfluidics for environmental analysis. Analyst 137:24–34
6. Sackmann EK, Fulton AL, Beebe DJ (2014) The present and future role of microfluidics in biomedical research. Nature 507:181–189
7. Ng AHC, Uddayasankar U, Wheeler AR (2010) Immunoassays in microfluidic systems. Anal Bioanal Chem 397:991–1007
8. Gervais L, de Rooij N, Delamarche E (2011) Microfluidic chips for point–of–care immunodiagnostics. Adv Mater 23:H151–H176
9. Lee H, Ham D, Westervelt RM (eds) (2007) CMOS biotechnology. Springer, New York
10. Iverson BD, Garimella SV (2008) Recent advances in microscale pumping technologies: a review and evaluation. Microfluid Nanofluidics 5:145–174
11. Mark D, Haeberle S, Roth G et al (2010) Microfluidic lab–on–a–chip platforms: requirements, characteristics and applications. Chem Soc Rev 39:1153–1182
12. Nge PN, Rogers CI, Woolley AT (2013) Advances in microfluidic materials, functions, integration, and applications. Chem Rev 113:2550–2583
13. Juncker D, Schmid H, Drechsler U et al (2002) Autonomous microfluidic capillary system. Anal Chem 74:6139–6144
14. Gervais L, Delamarche E (2009) Toward one–step point–of–care immunodiagnostics using capillary–driven microfluidics and PDMS substrates. Lab Chip 9:3330–3337
15. O'Farrell B (2009) Lateral flow immunoassay. Humana Press, New York
16. Wild D (2013) The immunoassay handbook. Elsevier, Oxford, UK
17. Lorenz H, Paratte L, Luthier R et al (1996) Low–cost technology for multilayer electroplated parts using laminated dry film resist. Sens Actuator A-Phys 53:364–368
18. Vulto P, Glade N, Altomare L et al (2005) Microfluidic channel fabrication in dry film resist for production and prototyping of hybrid chips. Lab Chip 5:158–162
19. Wangler N, Gutzweiler L, Kalkandjiev K et al (2011) High–resolution permanent photoresist laminate TMMF for sealed microfluidic structures in biological applications. J Micromech Microeng 21:095009
20. Temiz Y, Delamarche E (2014) "Chip–olate" and dry–film resists for efficient fabrication, singulation and sealing of microfluidic chips. J Micromech Microeng 24:097001
21. Ito T, Kawaguchi T, Miyoshi H et al (2007) Characterization of a microfluidic device fabricated using a photosensitive sheet. J Micromech Microeng 17:432–438
22. Zimmermann M, Schmid H, Hunziker P, Delamarche E (2007) Capillary pumps for autonomous capillary systems. Lab Chip 7:119–125

Chapter 3

Capillary-Driven Microfluidic Chips for Miniaturized Immunoassays: Patterning Capture Antibodies Using Microcontact Printing and Dry-Film Resists

Yuksel Temiz, Robert D. Lovchik, and Emmanuel Delamarche

Abstract

The miniaturization of immunoassays using microfluidic devices is attractive for many applications, but an important challenge remains the patterning of capture antibodies (cAbs) on the surface of microfluidic structures. Here, we describe how to pattern cAbs on planar poly(dimethylsiloxane) (PDMS) stamps and how to microcontact print the cAbs on a dry-film resist (DFR). DFRs are new types of photoresists having excellent chemical resistance and good mechanical, adhesive, and optical properties. Instead of being liquid photoresists, DFRs are thin layers that are easy to handle, cut, photo-pattern, and laminate over surfaces. We show how to perform a simple fluorescence immunoassay using anti-biotin cAbs patterned on a 50-μm-thick DF-1050 DFR, Atto 647N-biotin analytes, and capillary-driven chips fabricated in silicon.

Key words Microcontact printing, Patterning antibodies, Dry-film resist, Microfluidics, Immunoassay, PDMS

1 Introduction

Microfluidics are increasingly used for the analysis of samples in the context of research-driven experiments and clinical diagnostics [1]. A review by Manz et al. in 2002 already detailed well over 300 examples of applications in which microfluidic devices were used alone or with peripherals and "classical" analytical instruments [2]. A recent search of the literature indexed in "Web of Science," a database maintained by Thomson Reuters, returned over 30,000 articles, which had the term microfluidic(s) in their title. While being a very active field of research, a few bottlenecks remain with microfluidics. One of them is to seal microfluidic chips and connect them electrically and with liquids [3]. This topic is briefly covered in Chapter 3. Building on this Chapter, we address here a second bottleneck, which relates to the integration of protein receptors for assays performed in microfluidics [4]. Many assays require the immobilization of receptors such as cAbs, DNA probes, or ligands

Valérie Taly et al. (eds.), *Microchip Diagnostics: Methods and Protocols*, Methods in Molecular Biology, vol. 1547,
DOI 10.1007/978-1-4939-6734-6_3, © Springer Science+Business Media LLC 2017

for cells on the surface of microchannels. Taking the example of point-of-care immunodiagnostics, the patterning or localization of cAbs in well-defined areas of a device can be very challenging. The typical approaches for depositing proteins (i.e., cAbs) by adsorption from solution onto hydrophobic polystyrene surfaces [5] or nitrocellulose [6] are not adequate for functionalizing microchannels with cAbs. Instead, patterning cAbs in microfluidic channels relied on deposition from solution using stencils [7], local microcontact printing in recessed structures [8], patterning receptors using elastomeric microchannels [9, 10] or self-assembly using, e.g., DNA-directed immobilization [11]. Patterning cAbs into microfluidic structures using inkjet spotting [12] is similarly inadequate because it can lead to overfilling microstructures, adversary contamination of areas adjacent to receptor patterns, and drying artifacts. Here, we show how to use microcontact printing [13, 14] to pattern cAbs with high accuracy and contrast on a planar layer used to seal a capillary-driven microfluidic device. This method is general, simple, and efficient and builds on our earlier discovery that proteins can be microcontact printed from a PDMS stamp to surfaces [15, 16]. A monolayer of antibodies can be adsorbed from solution onto the hydrophobic surface of a micropatterned PDMS stamp. After rinsing and drying the stamp, proteins can be entirely transferred to a surface less hydrophobic than the stamp [17] while retaining significant biological activity [7]. We illustrate below how to prepare a planar PDMS stamp and ink it with a layer of cAbs [18]. Following a brief contact between the inked stamp and a hydrophilic micropatterned Si template, cAbs are removed in the areas of contact. The remaining patterned cAbs on the stamp are then microcontact printed on a DFR [19]. We recently showed how to use DFRs for sealing capillary-driven microfluidics and also that DFRs can be patterned with proteins [20]. Once patterned with proteins, the DFR is then laminated at low temperature on a microfluidic chip for sealing. In this example, ~50 × 50 μm^2 patterns of anti-biotin cAbs are used to capture Atto 647N-biotin analytes in PBS, leading to well-defined fluorescence signals measured at the surface of the DFR. Obviously, the types of receptors and analytes used, the patterns created, and assay conditions (volume of sample, flow rate, concentration of analyte) can be varied as appropriately needed for specific immunoassays.

2 Materials

Microfabrication of Si templates requires special equipment to perform photolithography on a Si wafer and etching of the structures. Although microfabrication can be outsourced, some technical lab equipment should be accessible. This involves a plasma chamber, a hot plate, compressed air, a standard oven, and a fluorescence microscope equipped with appropriate filter sets.

2.1 Fabrication of Si Template

1. Wafer: 4 inch (100 mm) single side polished Si wafer, N/Phos doping, 1–10 ohm.cm resistivity, <100> orientation, 525 ± 25 μm thickness (Si-Mat, Germany) (*see* **Note 1**).
2. Hexamethyldisilazane (Sigma-Aldrich GmbH, Switzerland).
3. Laboratory oven (e.g., Thermo Scientific Heraeus®, USA).
4. Photoresist: AZ®6612 (MicroChemicals GmbH, Germany) positive-tone photoresist.
5. Spin-coater: Standard spin-coating tool for 4 inch wafers.
6. Exposure tool: Direct laser writing tool DWL2000 (Heidelberg Instruments, Germany).
7. Photomask: 5 × 5 inch^2 soda lime glass substrate, Cr coating, 530 nm photoresist layer (Nanofilm, USA).
8. Photoresist developer: 20 % AZ®400K (MicroChemicals GmbH, Germany) potassium-based developer in deionized (DI) water.
9. DI water: Ultrapure water (Milli-Q, Merck & Cie, Switzerland) is used for all processes.
10. Deep reactive ion etching (DRIE): AMS 200 I-Speeder (Alcatel, France) using SF_6 gas for the etching cycle and C_4F_8 gas for the passivation cycle.
11. O_2 Plasma cleaning: GIGAbatch 310M resist ashing system (PVA TePla, Germany), O_2 flow: 500 sccm, power: 600 W, start pressure: 0.5 mbar.
12. Dicing: 8003 dicing saw (ESEC AG, Switzerland).
13. Acetone (Sigma-Aldrich GmbH, Switzerland).
14. IPA (2-propanol), Sigma-Aldrich GmbH, Switzerland).
15. Wafer box (Fluoroware, Entegris, USA).
16. Compressed air or N_2: filtered from dust particles and oil free.

2.2 Preparation of Planar PDMS Stamp

1. Petri dish 120 × 120 mm^2, 17 mm high (Greiner no. 688161, VWR Intern. GmbH, Switzerland).
2. PDMS: SYLGARD® 184 Silicone elastomer kit, mixed at a ratio of 1:10 (Dow Corning, Midland, MI, USA).
3. Disposable plastic container/beaker, 100–250 mL.
4. Balance with resolution of at least ±1 g (e.g., letter balance).
5. Water level.
6. Leveled oven for 60 °C overnight curing (SalvisLab Thermocenter, Renggli AG).
7. Scalpel or sharp utility knife.
8. Tweezers with flat and rounded tips.

2.3 Inking, Patterning Monolayer, and Printing on DFR

1. Glass slide: 75 × 25 × 1 mm^3 (Menzel, VWR International GmbH).
2. Petri dish: large enough to hold a glass slide (e.g., 90 mm diameter, PS, round, VWR International GmbH, Switzerland).
3. Pipettes: Set of Eppendorf Research® pipettes, 0.5 – 1000 µL (Vaudaux-Eppendorf AG, Germany).
4. PBS: Dulbecco's Phosphate Buffered Saline (Sigma-Aldrich GmbH).
5. cAb solution: anti-biotin antibodies (Abcam, UK), 250 µg/mL in PBS.
6. Dispensing bottle (e.g., product no. 215-8141, VWR International GmbH, Switzerland).
7. Microfluidic sealing layer (DFR): DF-1050 (Engineered Materials Systems Inc., USA), 50 µm thick; this is a high performance, chemically amplified, I-line sensitive, negative-tone DFR. The photosensitive adhesive layer is sandwiched between two plastic backing films. Store in the dark and preferably in a fridge (4–8 °C).
8. 2 mm diameter punch.
9. BSA solution: bovine serum albumin (Sigma-Aldrich GmbH), 1 % in PBS.

2.4 Chip Sealing and Performing an Immunoassay

1. Plastic spatula.
2. Plastic tweezers.
3. Atto 647N-biotin (Sigma-Aldrich GmbH, Switzerland): 1 µM solution in PBS.
4. Fluorescence microscope, Nikon Eclipse 90i (Nikon, Egg, Switzerland).

3 Methods

Microcontact printing is susceptible to dust particles and surface contaminations. It is preferable to perform as many steps as possible in a clean room or in a laminar flow hood. Perform all photolithography steps under yellow light, at room temperature and relative humidity of 40–50 %. Handle the DFR with gloves and dispose the excess DFR to the chemical waste according to the regulations.

3.1 Preparation of Si Templates (Fig. 1)

1. Take a 4 inch Si wafer. To enhance the photoresist adhesion, apply hexamethyldisilazane from the gas phase to the wafer surface for about 5 min while the wafer is heated at 110 °C.

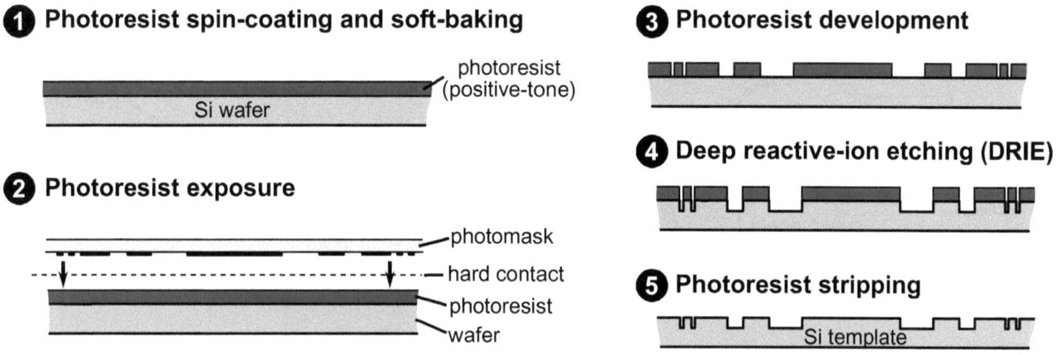

Fig. 1 Fabrication steps for Si template. (1) A Si wafer is spin-coated with photoresist and soft-baking is performed. (2) Photolithography and (3) photoresist development are performed to generate the template pattern in the layer of photoresist. (4) The pattern is transferred into the Si wafer using DRIE. (5) The residual photoresist is stripped from the fabricated Si template

2. Dispense about 1 mL AZ®6612 positive-tone photoresist on the wafer, spin-coat at 4000 rpm for 40 s, and bake the wafer at 110 °C for 1 min for a photoresist thickness of 1.2 μm.
3. Expose the wafer with an exposure tool, e.g., mask aligner or direct laser writing (40 mJ/cm^2 exposure dose), and develop the photoresist for 25 s in a glass beaker with 100 mL developer solution (mixture of 20 mL AZ®400K and 80 mL DI water).
4. Perform DRIE to etch the unprotected Si to a depth of 50 μm (*see* **Note 2**). Apply O_2 plasma cleaning for about 5 min to strip the photoresist and any remaining C_4F_8 passivation layer from the DRIE process.
5. Spin-coat a thin photoresist film to protect the wafer surface from dicing and bake at 110 °C for 1 min.
6. Dice the structured Si wafer to obtain single Si templates. Templates of 1–2 cm^2 in size are sufficient for microcontact printing and convenient for handling. Remove the photoresist in acetone and rinse the chips in IPA.
7. Store the templates in a clean container, e.g., Petri dish or wafer box, to protect them from dust particles until used. Shortly before a template is used for subtracting (*see* Subheading 3.3), clean the surface from potential organic contaminants and dust particles using an O_2-plasma (200 W, 30 s) and compressed air.

3.2 Preparation of Planar PDMS Stamp (Fig. 2)

1. Pre-position one or several Petri dishes of 120 × 120 mm^2 (*see* **Note 3**).
2. Mix PDMS base material with curing agent at a ratio of 10:1 thoroughly in a clean plastic beaker. The amount of PDMS required for one of the above mentioned Petri dishes is 20–25 g (*see* **Notes 4** and **5**).

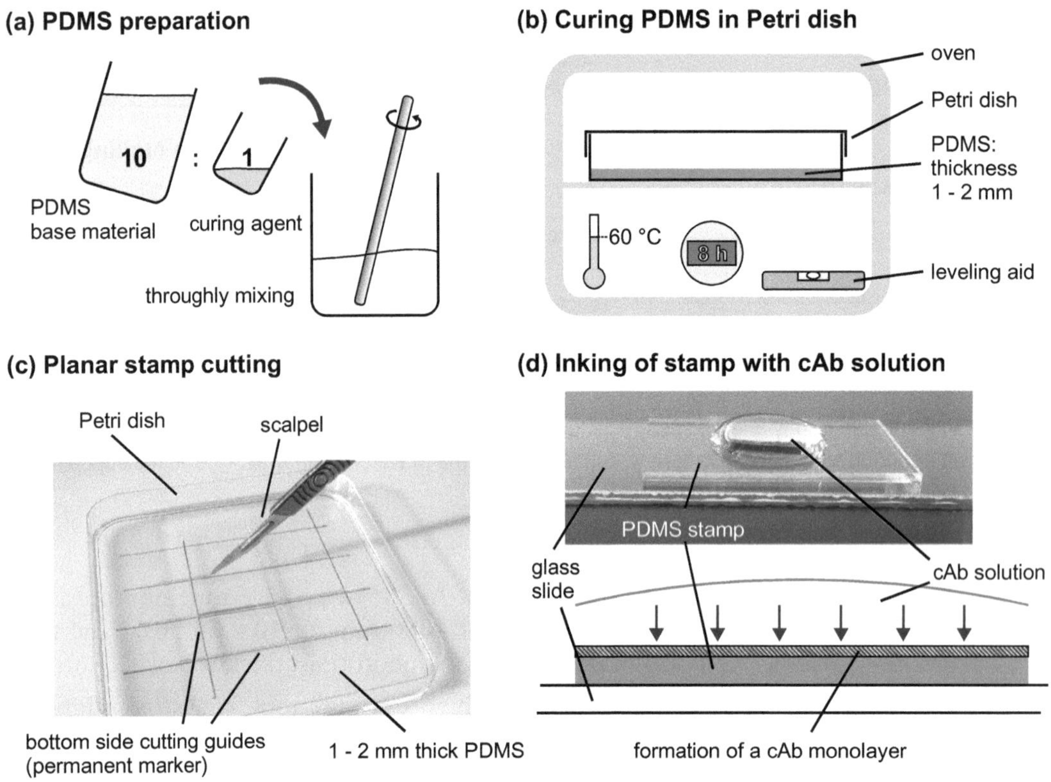

Fig. 2 Preparation of planar PDMS stamps and inking stamps with cAbs for microcontact printing. (**a**) The PDMS base material is thoroughly mixed with the curing agent at a ratio of 10:1 and poured into a Petri dish to a sheet thickness of 1–2 mm. (**b**) Curing of the PDMS takes place in a leveled oven at 60 °C, ideally overnight. (**c**) Individual planar stamps are cut from the PDMS sheet using a sharp knife. (**d**) A drop of cAb solution is placed on the active face of a stamp and spontaneous adsorption of cAbs from solution onto the PDMS leads to the formation of a monolayer of cAbs on the PDMS stamp

3. Place an open Petri dish on a balance and pour the PDMS into the dish while observing the weight.
4. Close the Petri dish and ensure that the PDMS covers the dishes bottom surface completely. Tilting may be necessary to help the PDMS spreading.
5. Temporarily place the Petri dish with the PDMS on a flat surface at RT until no air bubbles can be observed. This may take several hours.
6. Place the Petri dish in a leveled oven and let the PDMS cure at 60 °C for at least 8 h, typically overnight.
7. Use a sharp knife, e.g., scalpel, to cut planar stamps of the appropriate size from the PDMS sheet in the Petri dish. As a guide for cutting, lines can be drawn on the backside of the

dish using a permanent marker. The side of the PDMS that is in contact with the Petri dish will be the side used for printing and is referred to as the active face.

8. Cut PDMS directly in the Petri dish for storage and remove the individual stamps shortly before use with tweezers.

3.3 Patterning cAbs on DFR and Seal Chip with the DFR

1. Place a stamp with the active face up on a glass slide and place it into a clean Petri dish. Add about 1–2 mL of water to the Petri dish to ensure high humidity during inking and to prevent significant evaporation of the cAb solution. Apply cAb solution onto the stamp using a pipette. Approximately 100 μL of cAb solution is needed to cover about 1 cm^2 of the stamp. The tip of the pipette can be used to help spreading the cAb solution over the stamp. Close the Petri dish and incubate at RT for 30 min.
2. Remove the glass slide with the stamp from the Petri dish and rinse the stamp with 3–5 mL PBS and about 10 mL of purified water using a pipette or dispensing bottle. Gently dry with filtered compressed air or N_2.
3. A monolayer of cAbs should have formed on the PDMS during incubation, which now can be subtractive printed using a Si template. Therefore, place the active (inked) face of the stamp carefully onto a cleaned Si template. A conformal contact between the Si template and the active face of the stamp occurs within seconds (*see* **Note 6**). Leave the stamp in contact with the Si template for about 5 s, then separate the stamp from the Si template using tweezers (Fig. 3, **steps 1** and **2** and Fig. 4) (*see* **Notes** 7 and **8**).
4. Cut the DFR about to the size of the microfluidic chip using a scissor or scalpel. Punch a 2 mm diameter hole for the loading pad opening using a printed layout of the microfluidic chip for alignment (for multichip or wafer-level DFR processing, a cutting die or an electronic draft cutter can be used). Remove the protective film on one side of the DFR using Scotch tape or tweezers. Print the remaining antibody pattern on the stamp onto the DRF by placing the active face of the stamp onto the DFR and leave it in place for about 5 s. Remove the stamp using tweezers (Fig. 3, **step 3**).
5. Block the surface of the DFR by applying 1 % BSA solution for 10 min. Perform this step in a Petri dish with about 1–2 mL of water to avoid evaporation of the BSA solution.
6. Rinse the DFR with 3–5 mL PBS and about 10 mL of purified water. Gently dry with compressed air or N_2 (*see* **Note 9**).

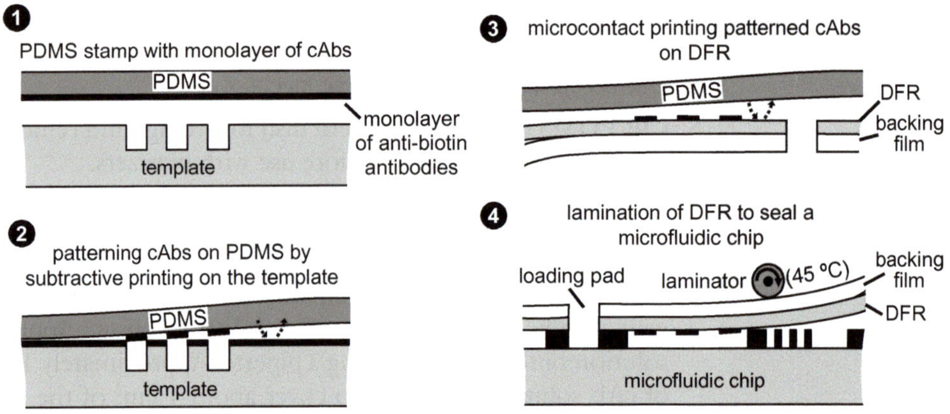

Fig. 3 Patterning capture antibodies on DFR used for sealing capillary-driven microfluidic chips. (1 and 2) A planar PDMS stamp with a monolayer of cAbs is carefully placed on a Si template and removed again after a 5 s to pattern the monolayer on the PDMS stamp. The resulting cAb pattern on the PDMS stamp is then transferred to a sheet of DFR (3) and the DFR laminated onto microfluidic chips (4) to seal the chip while at the same time integrating the cAb pattern for an immunoassay

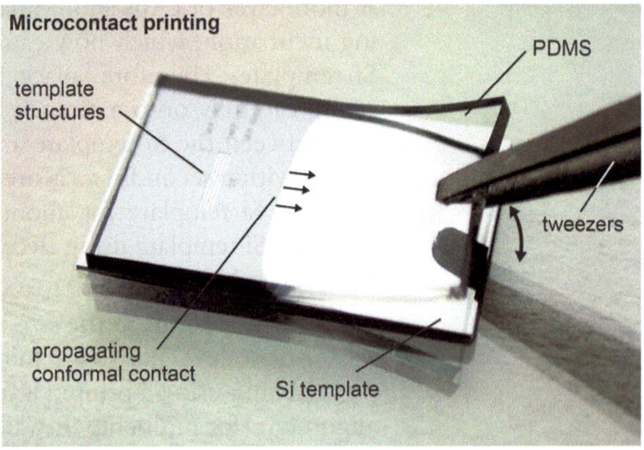

Fig. 4 Planar PDMS stamp in contact with a Si template. The monolayer of cAbs on the active face of a planar PDMS stamp is patterned by carefully placing the stamp on the top of a clean Si template. The stamp is slowly laid down on the Si template using tweezers. Propagation of the conformal contact between the two materials can be observed. After about 5 s, the tweezers are used to gently lift the stamp on one edge of the stamp and separate it from the Si template vertically

3.4 Chip Sealing and Performing the Immunoassay

1. Place the chip, comprising microfluidic structures, on a hot plate and set the hot plate temperature to 45 °C (*see* **Note 9**).
2. Align DFR over Si microfluidic chip and laminate the DFR to seal the chip by gently pressing with the help of a plastic

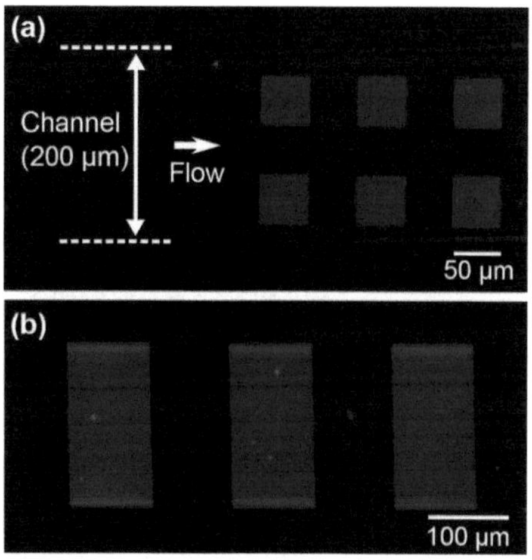

Fig. 5 Fluorescence microscope images showing an immunoassay performed on a DFR that has been patterned with cAbs using microcontact printing. Anti-biotin cAbs have been patterned on a DFR film used to seal a capillary-driven microfluidic chip. The addition of 2 µL of 1 µM Atto 647N-biotin in PBS to the chip resulted in the specific capture of Atto 647N-biotin in the areas where cAbs were patterned. Details on the design, fabrication, and flow properties of such a chip are provided in Chapter 2. The geometry of the signal areas, e.g., (**a**) squares or (**b**) rectangles are complementary to the pattern geometry of the Si template used to pattern the PDMS stamp

spatula or plastic wafer tweezers (Fig. 3, **step 4**). Laminated regions can be monitored by eye because they show a slight change in contrast (*see* **Notes 10** and **11**).

3. Apply 2 µL of a 1 µM solution Atto 647N-biotin (i.e., antigen) in PBS to the loading pad of a sealed chip. Subsequently add 4 µL of PBS to the loading pad.
4. Observe the capture of the Atto-647N-biotin by the anti-biotin cAb using a fluorescence microscope. Visible signals start to appear already after 1 min (*see* **Note 12**). The images in Fig. 5 demonstrate that (1) antibodies can be patterned by microcontact printing on DFR, (2) the antibodies retain functionality after the lamination step at 45 °C, and (3) the resolution and contrast of the fluorescent patterns on chips singulated using the "chip-olate" method [20] are excellent.

4 Notes

1. Wafer specifications (thickness, crystal orientation) are not critical for DRIE or microcontact printing. The polished side of the wafer is used.
2. 1.2 µm photoresist can be used to etch about 100 µm Si with an etch rate of 5 µm/min.

3. Square Petri dishes of 120 × 120 mm^2 are preferable, other sizes work as well, as long as the bottom of the dish is flat and smooth.
4. Generally, the resulting thickness of the PDMS poured into the Petri dish should be between 1 and 3 mm.
5. Wear safety glasses, gloves, and a lab coat when working with uncured PDMS. Especially the eyes need to be protected.
6. There is no need for applying additional pressure on the stamp. If an air pocket forms, slight pressing with tweezers may help, but lateral distortion of the stamp should be avoided to prevent deformation of the resulting antibody pattern.
7. The movement of pulling the stamp from the Si template should be smooth and perpendicularly to the surface plane. Lifting should start from one edge or corner only. A second pair of tweezers may be required to hold the Si template down.
8. Deposition of cAbs from solution onto PDMS is similar to coating cAbs to the polystyrene surface of microtiter wells and is in both cases spontaneous (driven by hydrophobic interactions between the cAbs and the surface) and self-limiting to the formation of a monolayer of cAb if time and the concentration of cAb in the ink are sufficient. The stamp cannot be reused after microcontact printing to a DFR.
9. DFR is a negative-tone photosensitive film, which crosslinks with UV light; therefore, long exposure to daylight should be avoided. We find that protein printing and BSA blocking steps performed in standard room light for about 30 min do not adversely affect the adhesion of DFR to the microfluidic chip.
10. In this example, we etched 50 μm deep microfluidic structures together with the Si template using DRIE. Alternatively, microfluidic chips can be patterned using SU-8 or DFR processing or by etching glass substrates. We find that DF-1050 has sufficiently good adhesion to Si, glass, SU-8, and other DFRs when used as a sealing layer.
11. The protective backing film on the top side of the DFR can be left as a part of the final device if the lamination is performed on a single chip. Multiple chips or the whole wafer can be laminated using a hot roll laminator after partially dicing the wafer containing the microfluidic chips. In this case, the backing film should be removed before singulating the chips by breaking, please see Chapter 2.
12. Different variants of assays can be performed using the techniques described here. Patterning cAbs on DFR and exposing them to labeled antigens using a capillary-driven microfluidic chip is just one example. A variant can be a "sandwich" immunoassay, where patterned cAbs are exposed to unlabeled

antigens, followed by a rinsing step and subsequent exposure to labeled detection antibodies (dAbs). In this case, each time a solution containing a different antibody or analyte is loaded to the microfluidic chip, a rinsing step would be required, similarly to what is done in ELISA assays using microtiter plates. Each of the steps above requires pipetting of 1–2 μL of solution onto the loading pad of the microfluidic chip.

References

1. Bange A, Halsall HB, Heineman WR (2005) Microfluidic immunosensor systems. Biosens Bioelectron 20:2488–2503
2. Auroux PA, Iossifidis D, Reyes DR et al (2002) Micro total analysis systems. 2. Analyticalstandard operations and applications. Anal Chem 74:2637–2652
3. Temiz Y, Lovchik RD, Kaigala GV et al (2015) Lab-on-a-chip devices: how to close and plug the lab? Microelectron Eng 132:156–175
4. Hitzbleck M, Delamarche E (2013) Reagents in microfluidics: an 'in' and 'out' challenge. Chem Soc Rev 42:8494–8516
5. Nakanishi K, Sakiyama T, Imamura K (2001) On the adsorption of proteins on solid surfaces, a common but very complicated phenomenon. J Biosci Bioeng 91:233–244
6. Hawkes R, Niday E, Gordon J (1982) A dot-immunobinding assay for monoclonal and other antibodies. Anal Biochem 119:142–147
7. Ziegler J, Zimmermann M, Hunziker P et al (2008) High-performance immunoassays based on through-stencil patterned antibodies and capillary systems. Anal Chem 80:1763–1769
8. Foley J, Schmid H, Stutz R et al (2005) Microcontact printing of proteins inside microstructures. Langmuir 21:11296–11303
9. Bernard A, Michel B, Delamarche E (2001) Micromosaic immunoassays. Anal Chem 73:8–12
10. Fan R, Vermesh O, Srivastava A et al (2008) Integrated barcode chips for rapid, multiplexed analysis of proteins in microliter quantities of blood. Nat Biotechnol 26:1373–1378
11. Niemeyer CM, Sano T, Smith CL et al (1994) Oligonucleotide directed self-assembly of proteins: semisynthetic DNA-streptavidin hybrid molecules as connectors for the generation of macroscopic arrays and the construction of supramolecular bioconjugates. Nucleic Acids Res 22:5530
12. Yatsushiro S, Akamine R, Yamamura S et al (2011) Quantitative analysis of serum procollagen type I C-terminal propeptide by immunoassay on microchip. PLoS One 6:e18807
13. Kumar A, Whitesides GM (1993) Features of gold having micrometer to centimeter dimensions can be formed through a combination of stamping with an elastomeric stamp and an alkanethiol "ink" followed by chemical etching. Appl Phys Lett 63:2002–2004
14. Xia Y, Whitesides GM (1998) Soft lithography. Angew Chem Int Ed 37:550–575
15. Bernard A, Delamarche E, Schmid H et al (1998) Printing patterns of proteins. Langmuir 14:2225–2229
16. Bernard A, Renault JP, Michel B et al (2000) Microcontact printing of proteins. Adv Mater 12:1067–1070
17. Tan JL, Tien J, Chen CS (2001) Microcontact printing of proteins on mixed self-assembled monolayers. Langmuir 18:519–523
18. Coyer SR, García AJ, Delamarche E (2007) Facile preparation of complex protein architectures with sub-100-nm resolution on surfaces. Angew Chem Int Ed 46:6837–6840
19. Vulto P, Glade N, Altomare L et al (2005) Microfluidic channel fabrication in dry film resist for production and prototyping of hybrid chips. Lab Chip 5:158–162
20. Temiz Y, Delamarche E (2014) 'Chip-olate' and dry-film resists for efficient fabrication, singulation and sealing of microfluidic chips. J Micromech Microeng 24:097001

Chapter 4

Fabrication and Evaluation of Microfluidic Immunoassay Devices with Antibody-Immobilized Microbeads Retained in Porous Hydrogel Micropillars

Toshihiro Kasama, Noritada Kaji, Manabu Tokeshi, and Yoshinobu Baba

Abstract

Due to the inherent characteristics including confinement of molecular diffusion and high surface-to-volume ratio, microfluidic device-based immunoassay has great advantages in cost, speed, sensitivity, and so on, compared with conventional techniques such as microtiter plate-based ELISA, latex agglutination method, and lateral flow immunochromatography. In this paper, we explain the detection of C-reactive protein as a model antigen by using our microfluidic immunoassay device, so-called immuno-pillar device. We describe in detail how we fabricated and used the immuno-pillar devices.

Key words Lab-on-a-chip, Microfluidic immunoassay, Biomarker

1 Introduction

Microfluidic devices have great potential as the biosensors, reactors, and so on. This is attributed to their inherent advantages including high surface-to-volume (S/V) ratio and reagent saving. Using microfluidic platforms, various biological [1, 2] as well as clinical studies [3] have been carried out. Especially, the use of microbeads in the microfluidic devices has attracted the special attention due to their ultra-high S/V ratio, whereas it has been a quite problem how to fasten the microbeads in the microchannels. To retain the microbeads in the microchannel, Sato et al. fabricated the dam structured microfluidic devices [4, 5]. They reported that the detection limit of the carcinoembryonic antigen (CEA) was 0.03 ng/mL. In addition, Kakuta et al. evaluated the reusability of the devices in quantitative analysis of brain natriuretic peptide and revealed their excellent reliability [6]. Rissin et al. performed ELISA on the microbeads, which were individually confined in water drops [7]. The detection limit of the prostate-specific antigen (PSA) was as low as 14 fg/mL. Recently, we proposed a novel

Valérie Taly et al. (eds.), *Microchip Diagnostics: Methods and Protocols*, Methods in Molecular Biology, vol. 1547,
DOI 10.1007/978-1-4939-6734-6_4, © Springer Science+Business Media LLC 2017

microfluidic immunoassay device, in which antibody-immobilized microbeads were retained in porous hydrogel micropillars so as to be high concentration of the capture antibodies in the microchannel. We called this the immuno-pillar device, with which biomarkers (a-fetoprotein [8], PSA [8], C-reactive protein (CRP) [8], monocyte chemotactic protein 1 [9], angiotensinogen [9], and liver-type fatty acid binding protein [9]) and staphylococcal enterotoxins A, B, C, D, and E [10, 11] were successfully detected with high sensitivities. Here, we explain the device fabrication and immunoassay procedures in detail while performing CRP immunoassay by using the immuno-pillar devices.

2 Materials

Prepare all solutions at room temperature in ambient light conditions.

1. Washing buffer: Phosphate buffered saline (PBS, pH 7.4) solution. Dissolve phosphate buffered saline powder (Wako Pure Chemical Industries, Osaka, Japan) in Millipore water (Millipore Corporation, Billerica, MA).
2. Antigen dilution buffer: The same buffer as for washing.
3. Coating buffer: Carbonate buffer (CB, pH 9.7) solution. Prepare with sodium carbonate and sodium hydrogen carbonate (Wako Pure Chemical Industries).
4. Detection antibody dilution buffer: The same buffer as for washing.
5. Polystyrene beads suspension: Polybead microspheres 1.00 μm (Polysciences Inc., Warrington, PA).
6. CRP solution: Human CRP (Scipac Ltd., Kent, UK).
7. Capture antibody: Goat antihuman CRP antibody affinity purified (A80-125A, Bethyl Laboratories, Inc., Montgomery, TX). Dilute to 1 mg/mL with coating buffer prior to use.
8. Detection antibody: Goat antihuman CRP antibody fluorescein isothiocyanate conjugated (A80-125F, Bethyl Laboratories, Inc.). Dilute to 50 mg/mL with detection antibody dilution buffer prior to use.
9. 1 % Bovine serum albumin (BSA) in PBS: Blocker BSA in PBS (10×) (Thermo Fisher Scientific Inc., Loughborough, UK) diluted tenfold in PBS.
10. Microtube rotator: MTR-103 (AS ONE, Osaka, Japan).
11. Polyethylene glycol based photocrosslinkable prepolymer (PP) solution: MI-1 (Kansai Paint Co., Ltd., Osaka, Japan).
12. Photoinitiator (PI) solution: PIR-1 (Kansai Paint Co., Ltd.).

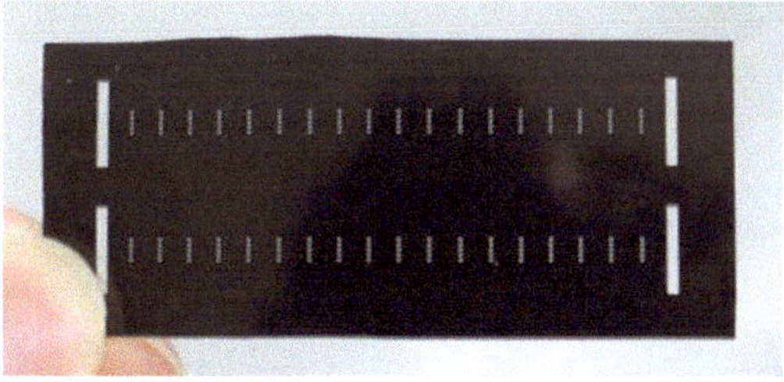

Fig. 1 Photomask for fabricating the immuno-pillar devices (3 cm × 7 cm)

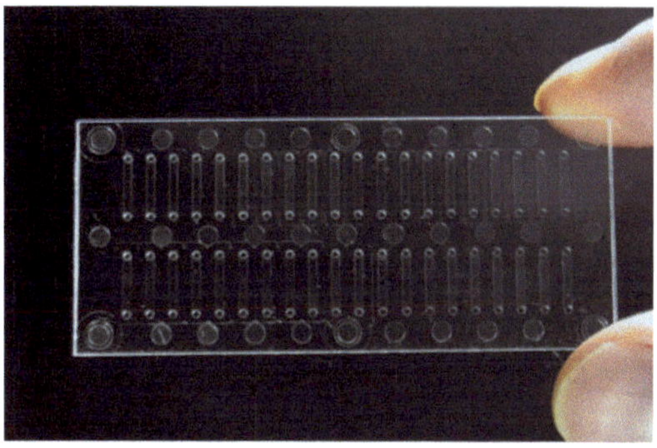

Fig. 2 Cyclic olefin polymer substrate of the immuno-pillar devices (3 cm ×7 cm) (Reproduced from [11] with permission from the Royal Society of Chemistry)

13. Mercury lamp: 20 mW (365 nm).
14. Micropipet: Use to inject the solutions into the microchannel. P-2 (PIPETMAN, Gilson S.A.S.).
15. Vacuum pump: Use to remove the solutions from the microchannel. Attach a pipet chip to the end of the tube. Catalogue number SP20, Air Liquide (Padova, Italy).
16. Photomask: Printed pattern on a transparency sheet (Institute of Microchemical Technology Co., Ltd., Kanagawa, Japan) (Fig. 1).
17. Fluorescence microscope: An inverted fluorescence microscope (IX-71, Olympus, Tokyo, Japan) equipped with a CCD camera (EB-CCD, Hamamatsu Photonics, Hamamatsu, Japan) and a laser (air-cooled Ar ion laser, l = 488 nm, 35LAL415, Melles Griot, Carlsbad, CA).
18. Cyclic olefin polymer (COP) substrate (Fig. 2): Two types of COP plates were fabricated by the injection molding method: flat plate (500 mm of thickness) and patterned plate (750 mm). Then, these COP plates were stuck each other to define the microchannels (Fig. 3). There were 40 microchannels on the substrate. Sumitomo Bakelite Co., Ltd. (Tokyo, Japan).
19. Analysis software: AquaCosmos, Hamamatsu Photonics, Hamamatsu, Japan.

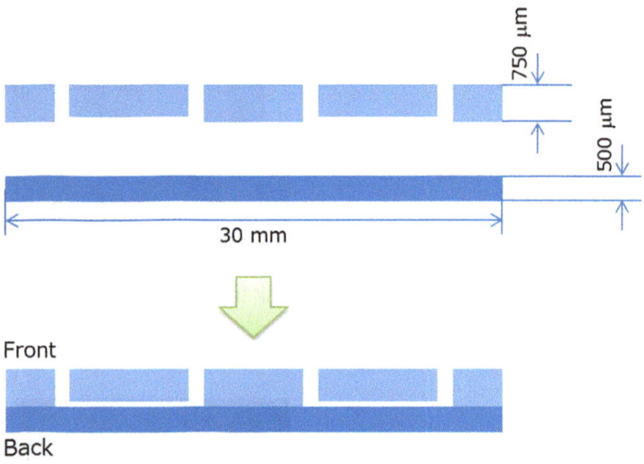

Fig. 3 Fabrication process of the microchannels in the COP substrate (cross section). The dimension of the microchannel is 1 mm in width, 40 mm in height, and 6.5 mm in length

3 Methods

3.1 Preparation of Antibody Immobilized Polystyrene Beads

Perform all procedures at room temperature in ambient light conditions unless otherwise specified. Follow waste disposal regulations when disposing reagents including sodium azide.

1. Dispense 100 mL of the polystyrene beads suspension in a 1.5 mL microtube (*see* **Note 1**).
2. Push the polystyrene beads in the suspension to the bottom of the microtube by centrifugation (20,000 × *g*, 1 min), and then remove the supernatant.
3. Add 1 mL of the antibody immobilization buffer and pipet up and down to resuspend the microbeads.
4. Push the polystyrene beads in the suspension to the bottom of the microtube by centrifugation (20,000 × *g*, 1 min), and then remove the supernatant.
5. Repeat **steps 3** and **4** three times.
6. Add 200 mL of the capture antibody (concentration) to the polystyrene beads and pipet up and down the suspension to resuspend the microbeads.
7. Stir at 5 rpm for 1 h by using the microtube rotator.
8. Incubate at 4 °C for at least 12 h.
9. Push the polystyrene beads in the suspension to the bottom of the microtube by centrifugation (20,000 × *g*, 1 min), and then remove the supernatant.

10. Add 1 mL of the washing buffer and pipet up and down to resuspend the microbeads.
11. Push the polystyrene beads in the suspension to the bottom of the microtube by centrifugation (20,000 × *g*, 1 min), and then remove the supernatant.
12. Repeat **steps 10** and **11** three times.
13. To prevent nonspecific binding, add 400 mL of 1 % BSA in PBS and incubate for 45 min.

3.2 Fabrication of Immuno-Pillar Devices

Perform all procedures at room temperature. To avoid an exposure to UV light, all processes must be carried out under yellow light.

1. Mix 3.2 mL of the PI solution, 32.3 mL of the Millipore water, and 64.5 mL of the PP solution in the other 1.5 mL microtube. Then, vortex the mixed solution (A solution) gently.
2. Add 100 mL of the antibody-immobilized polystyrene beads suspension to the A solution. Then, pipet up and down the mixed solution (AP solution).
3. Inject 0.25 mL of the AP solution into the microchannel.
4. Mount the photomask on the back of the COP substrate (*see* Fig. 3 and **Note 2**).
5. Irradiate UV light for ca. 10 s. The exposed areas become hydrogel pillars. These structures are called "immuno-pillars."
6. Remove the nonpolymerized AP solution by using the suction pump.
7. Flush the microchannel five times with 0.25 mL of the PBS.
8. Inject 0.25 mL of the 1 % BSA in PBS.
9. In order to prevent the nonspecific binding of the antigens and capture antibodies, immerse the microchannel including the immuno-pillars in the 1 % BSA in PBS until use for at least 1 h. Schematic of the immuno-pillar device is shown in Fig. 4.

3.3 Immunoassay Procedures

Perform all procedures at room temperature in ambient light conditions unless otherwise specified.

1. Remove the 1 % BSA in PBS from the microchannel by using the suction pump.
2. Inject 0.25 mL of the CRP solution.
3. Incubate for 1, 3, or 5 min.
4. Flush immuno-pillars three times with the PBS.
5. Inject 0.25 mL of the detection antibody.
6. Incubate for 1, 3, or 5 min in the dark.
7. Flush the immuno-pillars three times with the PBS.

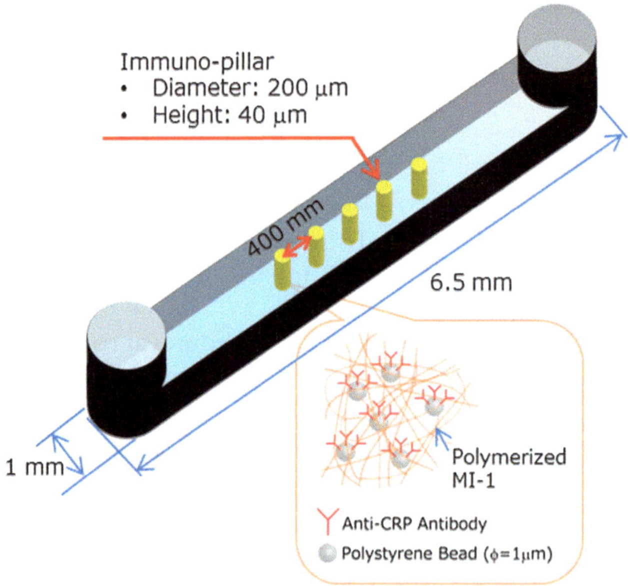

Fig. 4 Immuno-pillar device. Five hydrogel pillars (200 mm in diameter and 40 mm in height) including antibody-immobilized polystyrene beads were at the center of the microchannel with 400 mm of distance between neighboring pillars

8. Immerse the immuno-pillars in the PBS, and then observe fluorescence images of the immuno-pillars by using the fluorescence microscope (*see* **Note 3**).
9. Calculate the fluorescence intensity of the immuno-pillars with the analysis software (*see* **Note 4**).
10. Obtain the calibration curve as a result of tests with different concentrations (Fig. 5).

4 Notes

1. Microbeads tend to sink quickly, providing nonuniform dispersion. Dispense the microbeads suspension immediately after vortexing.
2. This is to protect the photomask against the adhesion of the AP solution.
3. Adjust the gain of the detector by using CRP solution with the highest concentration.
4. Select whole immuno-pillar as a region of interest when analyzing the fluorescence intensity.

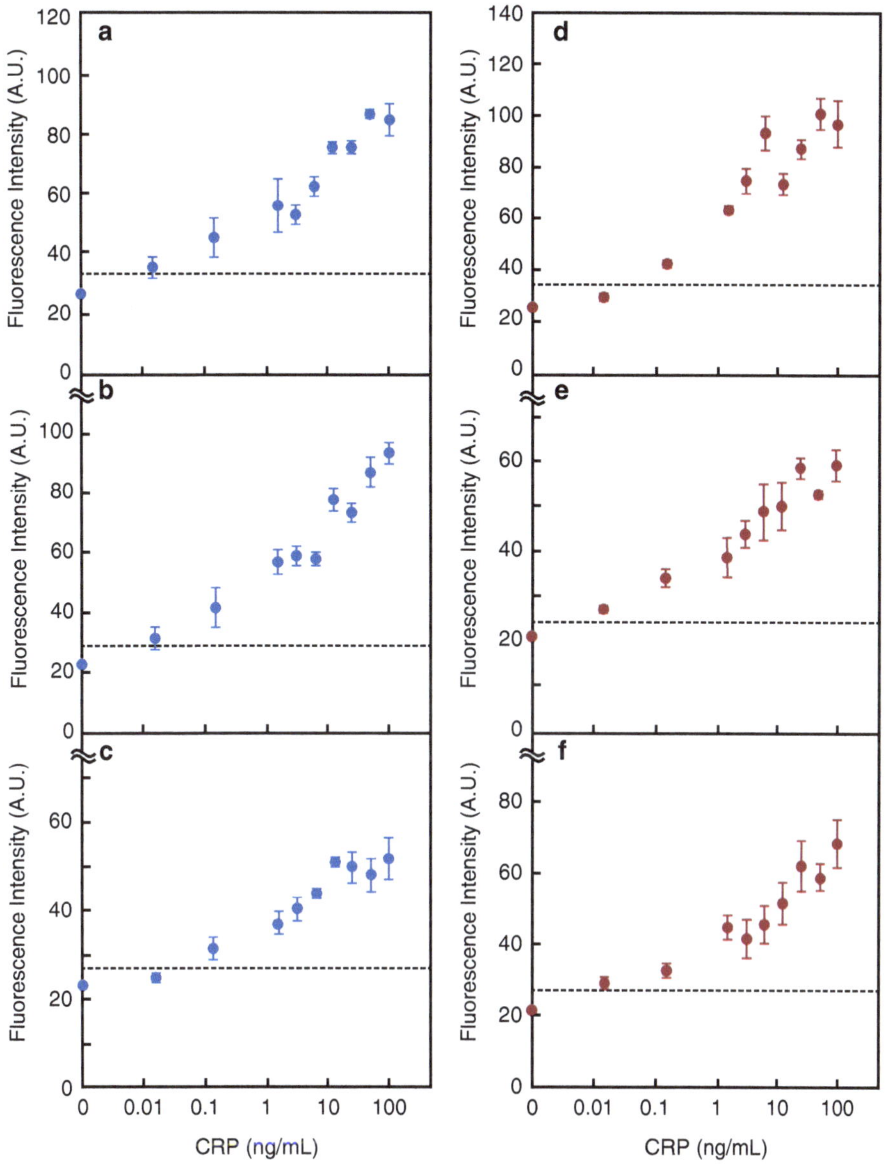

Fig. 5 Calibration curves of CRP. (**a**) 5 min incubation (standard samples). (**b**) 3 min incubation (standard samples). (**c**) 1 min incubation (standard samples). (**d**) 5 min incubation (serum samples). (**e**) 3 min incubation (serum samples). (**f**) 1 min incubation (serum samples). The *dashed line* represents the signal level at 3 standard deviations above the background (Reproduced from [8] with permission from the Royal Society of Chemistry)

Acknowledgment

This work was partially supported by the Japan Society for the Promotion of Science (JSPS, grant no. 24245013). We would like to thank Kansai Paint Co., Ltd., for valuable discussions.

References

1. Li P, Mao Z, Peng Z, Zhou L, Chen Y, Huang PH, Truica CI, Drabick JJ, El-Deiry WS, Dao M, Suresh S, Huang TJ (2015) Acoustic separation of circulating tumor cells. Proc Natl Acad Sci USA 112:4970–4975
2. Yasaki H, Onoshima D, Yasui T, Yukawa H, Kaji N, Baba Y (2015) Microfluidic transfer of liquid interface for parallel stretching and stamping of terminal-unmodified single DNA molecules in zigzag-shaped microgrooves. Lab Chip 15:135–140
3. Gervais L, de Rooij N, Delamarche E (2011) Microfluidic chips for point-of-care immunodiagnostics. Adv Mater 23:H151–H176
4. Sato K, Tokeshi M, Kimura H, Kitamori T (2001) Determination of carcinoembryonic antigen in human sera by integrated bead-bed immunoasay in a microchip for cancer diagnosis. Anal Chem 73:1213–1218
5. Sato K, Tokeshi M, Odake T, Kimura H, Ooi T, Nakao M, Kitamori T (2000) Integration of an immunosorbent assay system:analysis of secretory human Immunoglobulin A on polystyrene beads in a microchip. Anal Chem 72:1144–1147
6. Kakuta M, Takahashi H, Kazuno S, Murayama K, Ueno T, Tokeshi M (2006) Development of the microchip-based repeatable immunoassay system for clinical diagnosis. Meas Sci Technol 17:3189–3194
7. Rissin DM, Kan CW, Campbell TG, Howes SC, Fournier DR, Song L, Piech T, Patel PP, Chang L, Rivnak AJ, Ferrell EP, Randall JD, Provuncher GK, Walt DR, Duffy DC (2010) Single-molecule enzyme-linked immunosorbent assay detects serum proteins at subfemtomolar concentrations. Nat Biotechnol 28:595–599
8. Ikami M, Kawakami A, Kakuta M, Okamoto Y, Kaji N, Tokeshi M, Baba Y (2010) Immunopillar chip: a new platform for rapid and easy-to-use immunoassay. Lab Chip 10: 3335–3340
9. Sun M, Kasama T, Yuzawa Y, Akiyama S, Matsuo S, Kaji N, Tokeshi M, Baba Y (2012) Rapid urine-based clinical diagnosis of diabetic nephropathy with femto-molar sensitivity by immunopillar devices. In: Proceedings of the micro total analysis system, Okinawa, Japan, pp 812–814
10. Jin W, Yamada K, Ikami M, Kaji N, Tokeshi M, Atsumi Y, Mizutani M, Murai A, Okamoto A, Namikawa T, Baba Y, Ohta M (2013) Application of IgY to sandwich enzyme-linked immunosorbent assays, lateral flow devices, and immunopillar chips for detecting staphylococcal enterotoxins in milk and dairy products. J Microbiol Methods 92:323–331
11. Kasama T, Ikami M, Jin W, Yamada K, Kaji N, Atsumi Y, Mizutani M, Murai A, Okamoto A, Namikawa T, Ohta M, Tokeshi M, Baba Y (2015) Rapid, highly sensitive, and simultaneous detection of staphylococcal enterotoxins in milk by using immuno-pillar devices. Anal Methods 7:5092–5095

Chapter 5

Using Electrophoretic Immunoassay to Monitor Hormone Secretion

Shusheng Lu and Robert T Kennedy

Abstract

It has been demonstrated that microfluidic systems allow integration of sampling, reagent mixing, and rapid electrophoretic analysis. They have also proven useful for culturing cells wherein control over the environment allows novel and automated experiments. Here, we describe a microchip-based electrophoresis assay that allows cell culture and hormone monitoring. An online gradient generator can control cell culture condition precisely. This system has been applied for Pancreas islets' glucose sensitivity studies.

Key words Microfluidics, Perfusion culture, Capillary electrophoresis, Competitive immunoassay, Islets of Langerhans, Diabetes, Glucose sensitivity

1 Introduction

Maintaining cells in a stable life-supporting environment is essential for cellular physiology studies. Traditionally, these studies are usually performed in a static physiological medium or perfused with a macroscale perfusion system [1–3]. Microfluidic techniques enable precise control of fluids and particles at nanoliter scale and integration of analytical functions, which make microfluidics a versatile platform for growing, manipulating, and chemical monitoring of living cells [4]. Some recent examples of microfluidic works are high-throughput automatic cell culture [5], monitoring single cell cytokine production in response to external stimuli [6], and the capture and chemical activation of single cells [7].

Secretion of chemical products from cells, which includes signal molecules (hormones or neurotransmitters), tropic factors, and metabolic products, is another important physiological function to study. Perfusing cells, collecting fractions, and performing off-line analysis by immunoassay or other appropriate methods are usually applied to measure temporally resolved cell secretory [8]. The development of a microfluidic device combining cell culture and dynamic chemical monitoring would miniaturize and automate

Valérie Taly et al. (eds.), *Microchip Diagnostics: Methods and Protocols*, Methods in Molecular Biology, vol. 1547,
DOI 10.1007/978-1-4939-6734-6_5, © Springer Science+Business Media LLC 2017

such applications, and could become a valuable tool for cell physiological studies. Additionally, a lot of microfluidic cell culture systems combine gradients generators of stimulus (like drug) concentrations [9], to create better microenvironments.

In this chapter, we describe a microfluidic device that can be used to perfuse a cluster of cells and monitor secretion at 8 s intervals using a rapid electrophoresis-based competitive immunoassay [10]. Cells are continuously perfused with buffer on chip. A small fraction of perfusate containing cell secreta is continuously sampled by electroosmotic flow (EOF) and assayed for hormone.

The system is applied to monitoring insulin secretion from single islets of Langerhans. Islets are 75–200 μm diameter spheroid cell clusters located in the pancreas that contains 2000–4000 cells each, 70–80 % of which are insulin secreting β-cells. β-cells are stimulated to secrete insulin at elevated blood glucose concentration, which is crucial for maintaining glucose homeostasis [11]. Insulin secretion has complex dynamics. In vitro step increases in glucose concentration (typically from 3 to 11 mM) usually result in an initial burst (first phase) and oscillations with periods of 3–5 min (second phase) insulin secretions [12, 13]. Some evidence also suggests a secondary ultradian [14] and possible circadian [15] rhythm with 2 and 24 h periods. Deficiencies in these rhythms have been linked to type 2 diabetes (T2D).

Our laboratory previously developed a microfluidic device that can detect insulin secreted from single islets using an electrophoresis-based competitive immunoassay [16]. This device has been further developed for high-throughput parallel operation [17] and 24 h continuous monitoring [18]. Although these works demonstrated feasibility of microfabricated devices for cell culture and chemical monitoring, constant stimulus without precise, rapid, and automatic fluid control has some limitations: (1) Constant glucose cannot mimic real glucose fluctuation in animal bodies [19]. (2) Rapid stimulus change (e.g., glucose ramp) cannot be performed. In this work, we describe a modification of the chip that enables gradients generation.

The gradient generator implemented on chip enables glucose sensitivity studies, where glucose ramp is required [20]. The threshold glucose concentration where islets start secreting insulin when they are perfused with gradually ramping up glucose is an important factor to measure glucose sensitivity.

2 Materials

All solutions are prepared with Milli-Q (Millipore, Bedford, MA) 18 MΩ deionized water and filtered with 0.2 μm nylon syringe filters before using.

2.1 Chemicals and Reagents

1. Balance salt solution (BSS): 125 mM NaCl, 5.9 mM KCl, 1.2 mM $MgCl_2$, 2.4 mM $CaCl_2$, 25 mM Tricine, and 0.7 mg mL^{-1} bovine serum albumin (BSA). Prepare 1 L such solution without BSA and adjust to pH 7.4 with NaOH (*see* **Note 1**). Store at 4°C. Take 100 mL stock solution and add 70 mg BSA for daily use (*see* **Note 2**).
2. Immunoassay reagent buffer: 60 mM NaCl, 1 mM EDTA, 20 mM Tricine, 0.1 % (w/v) Tween 20, and 0.7 mg mL^{-1} BSA. Prepare 500 mL such solution without BSA and Tween 20, adjust to pH 7.4. Store at 4°C. Take 100 mL solution and add 70 mg BSA and 0.1 g Tween 20 for daily use.
3. Electrophoresis buffer: 20 mM NaCl and 150 mM Tricine. Prepare 200 mL such solution and adjust to pH 7.4.
4. Stock fluorescein isothiocyanate-labeled insulin (FITC-insulin) from Sigma (St. Louis, MO) is diluted to 166 μM in immunoassay reagent buffer and stored at −20°C (*see* **Note 3**). Stock monoclonal antibody (Ab) to human insulin from Biodesign International (Saco, ME) is stored at 4°C in the manufacturer-provided phosphate-buffered saline.
5. RPMI-1640 cell culture media contains 10 % fetal bovine serum, 100 U/mL of penicillin, and 100 μg/mL of streptomycin. pH is 7.4 at 37°C and 5 % CO_2.

2.2 Perfusion Culture

1. Fusion Touch 400 syringe pumps × 2 (Chemyx Inc., Stafford, TX).
2. Gastight 1700 series syringes (Hamilton Robotics Inc., Reno, NV; *see* **Note 4**).
3. Reducing union, 1/16″ to 360 μm (Vici Valco Instruments, Houston, TX; *see* **Note 5**).

3 Methods

3.1 Microfluidic Chip Fabrication and Preparation

Chip exposure, development, etching, Piranha, and RCA rinse and bonding are all carried in clean room. The chip is made out of glass.

1. 1 mm thick Borofloat photomask blanks (3.8 × 7.6 cm) coated with a layer of chrome and photoresist (Telic Company, Valencia, CA) are exposed to UV light for 6 s through a patterned photomask (Fineline Imaging Inc., Colorado Springs, CO; *see* **Note 6**). The exposed photomasks are developed in AZ 726 MIF developer (Clariant Corp., Summerville, NJ) for 30 s, followed by 2 min (*see* **Note 7**) developing in CEP-200 chrome etchant (Microchrome Technologies, Inc., San Jose, CA).

2. The exposed glass is etched in 17:24:79 (v/v/v) $HNO_3/HF/H_2O$ for 25 min resulting in 15 μm deep and 40 μm wide electrophoresis and 90 μm wide perfusion channels (*see* **Note 8**).
3. Fluid access holes are made in the etched glass with 360 μm drilling bits (*see* **Note 9**).
4. The remaining photoresist is removed with acetone (*see* **Note 10**), and the remaining chrome is removed with the CEP-200 chrome etchant.
5. The resulting chips and blank coverplates are placed in piranha solution (3:1, v/v, H_2SO_4/H_2O_2) (*see* **Note 11**) for 20 min and then RCA solution (5:1:1, v/v/v, $H_2O/NH_4/H_2O_2$) for at least 40 min (*see* **Note 12**).
6. The etched chips and blank coverplates are bonded at 640°C using a programmable furnace (*see* **Note 13**).
7. Microfluidic reservoirs (IDEX Health & Science, Oak Harbor, WA) are applied to the device over drilled access holes with Epoxy Glue. 80 μL reservoirs (IDEX N-131) are used for all ports except the perfusion inlets, which use 1/32″ tubing interconnect ports (IDEX N-124H, *see* **Note 14**).

3.2 Microfluidic Chip Operation

1. The layout of the channels of the microfluidic chip and diagrams of chip operation are shown in Fig. 1. The system is improved compared to a previous design by incorporating a method to control the glucose applied to the cells.
2. The chip is conditioned prior to experiments by flowing 0.1 M NaOH through the channels, followed by water and then experimental solutions (*see* **Note 15**).
3. All solutions are filtered before using to prevent introduction of particulates to the chip.
4. During the experiment, single islets are perfused with BSS containing glucose at 0.6 μL mL^{-1} (*see* **Note 16**).
5. For control of glucose, the system incorporates two independent syringe pumps loaded with low and high concentrations of glucose. Flow rates of both are controlled by a home-made LabView program with a total of 0.6 μL min^{-1}. By mixing two syringe pumps' contents at different flow rates through a Tee fabricated on the electrophoresis chip, arbitrary glucose concentrations between low and high glucose can be generated.
6. The glucose ramp is mimicked by fluorescein to verify stable changes as shown in Fig. 2.
7. Besides islet reservoir, to provide fresh buffer to all the other reservoirs, appropriate solutions are continuously pumped onto the chip via fused-silica capillaries inserted into vials in high pressure bomb, pressured with 12 psi of helium.

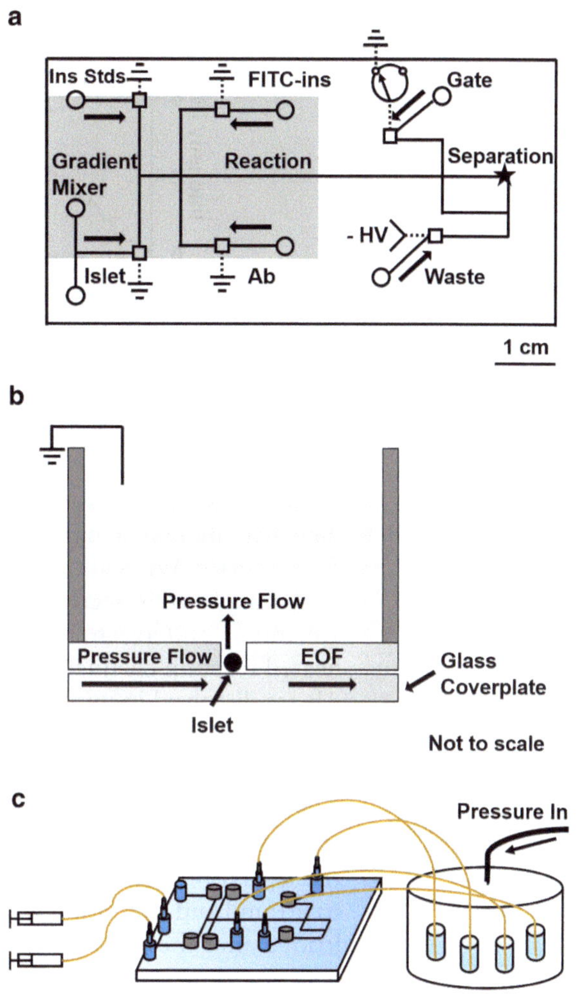

Fig. 1 A microfluidic chip for monitoring insulin secretion from single islets. (**a**) The channel layout of the microfluidic device. Microfluidic channels are indicated by *solid lines*; Electrical connections are indicated by *dotted lines*. *Circles* indicate perfusion inlets and *squares* indicate reservoirs that are sampled by electroosmotic flow (EOF). *Arrows* indicate directions of perfusion flow. *Gradient mixer* indicates the Tee fabricated on chip to generate arbitrary glucose gradients. The *shaded* portion is heated at 37°C with a thin film heater. The *star* indicates laser-induced fluorescence occurred 1 cm past injection point. –HV are where high voltage is applied. (**b**) *Side*—view of an islet perfusion chamber. Islet is loaded into a 0.1 μL chamber and perfused with pressure-driven flow, after which the fluid flows into a 100 μL fluidic reservoir. Solution with insulin from the chamber is sampled by EOF through the sampling channel at a rate of approximately 2 nL min^{-1}. (**c**) Scheme for perfusing reservoirs. Perfusion buffer supply is pushed by syringe pumps or pressurized and fluid flows into the chip to constantly supply buffer that is sampled by EOF

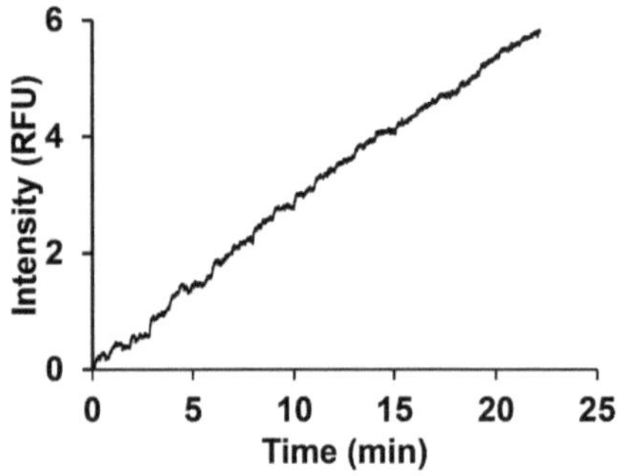

Fig. 2 Program controlled fast step change from 0 to 2 nM fluorescein. By changing the syringe pumps flow rates slightly each minute, mixed concentration gradually increases in the total 20 min time

8. Negative high voltage (−6 kV) is applied at the waste reservoir of the device. With all the other reservoirs grounded, solutions containing insulin secreted by islet and 100 nM FITC-insulin are mixed and reacted competitively with 50 nM Ab on a heated reaction channel (*see* **Note 17**). Both FITC-insulin and Ab are dissolved in immunoassay reagent buffer.
9. Electrophoresis buffer is pumped into gate and waste reservoirs. The gate reservoir is connected to ground via a high voltage relay. Injection is performed by opening the relay [21]. When the relay is opened, the sample is allowed to load onto the separation channel (*see* **Note 18**). The gate is then returned to ground and separation is performed. Another sample can be loaded after the previous separation is completed. The sample injection time is 0.5 s applied at 7.5 s intervals.
10. Laser-induced fluorescence detection occurs 1 cm from the injection cross. Each separation results in a FITC-insulin bound to antibody (B) peak followed by a free FITC-insulin (F) peak. Concentrations of insulin are quantified by comparing B/F peak areas (*see* **Note 19**) ratios of each electropherogram to a calibration curve.
11. Ground is changed from the islet reservoir to insulin standards reservoir for calibration.

3.3 Instruments and LIF Detection

1. LIF detection is performed with a Zeiss Axiovert 35 M inverted microscope equipped with a Photon Technology International 814 photometer (Birmingham, NJ).
2. The 488 nm Excitation light of a 20 mW optically pumped semiconductor Sapphire laser (Coherent, Santa Clara, CA) is directed onto a 500 nm long-pass dichroic mirror and through a 40×, 0.6 numerical aperture, long working distance objective (Carl Zeiss, Inc., Thornwood, NY).

3. After passing through the dichroic mirror, the emission light is further filtered through a 530 ± 30 nm band-pass filter. The fluorescence emission is further spatially filtered by an iris diaphragm on the photometer.
4. Instrument control and data collection are performed using LabVIEW software written in house (National Instruments, Austin, TX). High-throughput analysis of collected electropherograms is performed using Cutter software [22].

3.4 Islet Isolation Protocol

Pancreatic islets are obtained from 20 to 30 g male CD-1 mice. The method has previously been described [23].

1. Mice are sacrificed by cervical dislocation, and collagenase type XI is injected into the pancreas through the main pancreatic duct. The pancreas is removed and incubated in 5 mL of collagenase solution at 37°C for ~14 min (*see* **Note 20**).
2. A Ficoll gradient is used to separate exocrine tissue from endocrine tissue, and islets are picked up by hand under a stereomicroscope. Islets with oblong to spherical shape, 100–200 μm diameter, and intact membrane (showing smooth surface) are selected for experiments.
3. The islets are placed in RPMI-1640 cell culture media supplemented with 10 % fetal bovine serum, 100 unit mL^{-1} penicillin, and 100 μg mL^{-1} streptomycin at 37°C, 5 % CO_2, pH 7.4. Islets are used 2–5 days following isolation (*see* **Note 21**).

3.5 Application to Glucose Sensitivity Studies

1. One and 25 mM glucose are loaded on the two independent syringe pumps. Flow rates of both pumps are adjusted every minute to generate a glucose ramp from 1 to 25 mM with a rate of 1 mM min^{-1}.
2. Islets preincubated overnight with 3 mM glucose, 11 mM glucose, and 11 mM glucose + 200 nM diazoxide (blocking insulin secretion, Dz) are measured for insulin secretion. Islets have different threshold glucose when exposed to different culture conditions as shown in Fig. 3 (*see* **Note 22**).

4 Notes

1. Saturated NaOH can be used first to approach required pH faster. Adding less volume of NaOH also helps maintaining concentration of salts. However, to avoid sudden rise of pH passing required value, lower concentrations of NaOH (6–1 M) would be better to be used when getting close to the required pH.
2. To prepare 100 mL aliquot and avoid BSA (and Tween 20 in immunoassay reagent buffer) in stock solutions both help

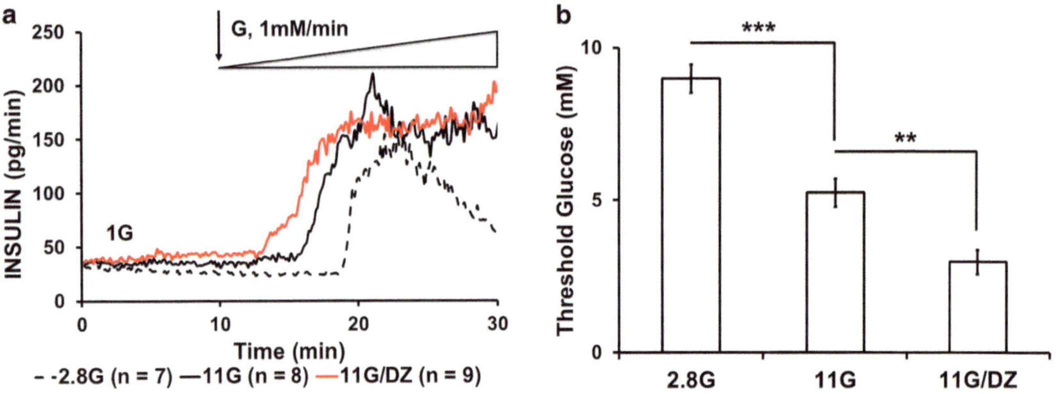

Fig. 3 Insulin secretory dynamics in the three treatment groups as measured using microfluidics technology. (**a**) Secretion shown is from mouse islets chronically exposed to 11 mM glucose (*solid black line*), 3 mM glucose (*dashed black line*), or 11 mM glucose + Dz (*redline*). The ordinate depicts average insulin secretion rate, in pg/min, of 7–9 single islets in each condition. The *triangle wedge* indicates applying of glucose ramp at the increasing rate of 1 mM/min, from 1 to 21 mM. (**b**) The bar graph showing "Threshold Glucose" of islets overnight cultured in 3, 11, or 11 mM glucose + Dz. *** $p < 0.001$, ** $p < 0.01$

keeping the solutions from degradation. One typical microfluidic experiment consumes 10 mL BSS, 4 mL immunoassay reagent buffer, and 2 mL electrophoresis buffer. The volume of each solution (1 L, 500 mL, and 200 mL) is prepared proportional to it. Each solution can be used for as long as 4 months ideally; however, degradation would be reflected by broadened or weak shapes of electrophoregram peaks, which indicates fresh solutions should be prepared.

3. The stock FITC-insulin solution is split into several small vials with 20 μL each to prevent frequent freeze and thaw.
4. If a longer period of experiment is required, larger sizes of syringes should be selected (e.g., Gastight 1000 series).
5. The reduction unions connect syringe to 360 diameter fused silica capillaries, which introduce the solution to the chip.
6. Patterns of microchannels are drawn in AutoCAD (Autodesk, Inc., San Rafael, CA). The photomask is prepress films with pattern (microfluidic channels) printed with width as small as 10 μm, which is the lowest resolution of printed photomasks. Should smaller resolution is required, emulsion (5 μm) or chrome (1 μm) photomasks can be applied. Perfusion channels are drawn 60 μm on photomask when electrophoresis channels are only 10 μm. Bigger perfusion channel helps in preventing particles clogging.
7. After chrome etching, patterns should be examined under the microscope that all desired parts have been well exposed. More etching time can be added if chrome is not fully removed.

8. Because HF is extremely corrosive and toxic, the standard operation procedure should be strictly followed. Briefly, prepare and use HF with a plastic container in a fume hood, and appropriate protective wears (long-sleeve HF-resistant gloves, face mask or goggles, etc.) should be applied. The glass etching rate with the provided recipe is 0.6 μL min^{-1}. Channel etching would expand to both sides (15 μm each) of the channels to make a reversed trapezoid shape.
9. After drilling, chips are rinsed and sonicated in water for 1 h before being brought back to clean room.
10. Ethanol also removes photoresist but with a slower rate. It is also less toxic than acetone.
11. *Piranha solution is corrosive and reacts violently with organic materials, particularly when hot. Standard operation procedure should be followed. It should not be stored in closed containers.*
12. RCA is prepared together with piranha and heated to boil before use.
13. After RCA, glasses are rinsed with water and examined under the microscope for any visible particles or dusts (placing back to RCA for a few minutes or rubbing with soap water help in removing them). Then etch chip and blank coverplates are bonded under flowing water. Make sure no air bubble exists between them. The bonded chips are carefully placed into a 70–90°C oven for 1–2 h, and checked without newton rings or water plugs in channels before putting in 640°C furnace.
14. To help in aligning the ports' holes with the fluid accessing holes on the chip, a 360 μm capillary is inserted through both of them and kept before epoxy is dry.
15. NaOH is usually flowed for 20 min. Remove all the NaOH with water before applying experimental solutions, since mixing of NaOH with experimental solutions can cause irreversible clogs in channels.
16. Before being loaded on chip, islets are transferred to a petri dish containing BSS. All transfers are done with a 10 μL pipette.
17. This portion of chip is heated with a thin-film resistive heater (Minco Products, Inc., Fridley, MN). The temperature of the islet chamber is kept at 37°C.
18. Sometimes, insufficient sample plug can be injected into the separation channel. Under this circumstance, the high voltage relay connected to gate reservoir can be thrown to another high voltage (−5 kV) to pull more sample for separation. A single pole double throw (SPDT) high voltage relay is thus required for this application.
19. We found comparing peak areas has less variation than peak heights.

20. The time of digestion varies among batches of collagenase and performance of collagenase injection. Collagenase P from Roche Diagnostics Corporation (Indianapolis, IN) is recommended for relatively stable activity. Nevertheless, the vial containing pancreas should be frequently checked after 10 min and quenched as soon as mixture being cloudy and very few large masses of pancreatic tissue remaining.
21. The best time to apply islets for physiological studies is still under debate. To the best of our knowledge, the best time would be 2–5 days after isolation (sometimes the 6th day is still acceptable). But after this period, islets lost their insulin secretion.
22. β-cells respond to chronic hyperglycemia by increasing their mass (increasing β-cell number and/or size) and secretory function (more glucose sensitive) in compensating for insulin resistance. Both insulin and glucose are candidate signals for triggering β-cell compensation, as both are elevated when there is insulin resistance. We separate the effects of insulin from those of glucose by chronically blocking insulin secretion using diazoxide along with glucose.

References

1. Bergsten P, Grapengiesser E, Gylfe E et al (1994) Synchronous oscillations of cytoplasmic Ca^{2+} and insulin release in glucose-stimulated pancreatic islets. J Biol Chem 269:8749–8753
2. Liu Y-J, Tengholm A, Grapengiesser E et al (1998) Origin of slow and fast oscillations of Ca^{2+} in mouse pancreatic islets. J Physiol 508(2):471–481
3. Zraika S, Dunlop M, Proietto J, Andrikopoulos S (2002) The hexosamine biosynthesis pathway regulates insulin secretion via protein glycosylation in mouse islets. Arch Biochem Biophys 405:275–279
4. Mehling M, Tay S (2014) Microfluidic cell culture. Curr Opin Biotechnol 25:95–102
5. Gómez-Sjöberg R, Leyrat AA, Pirone DM et al (2007) Versatile, fully automated, microfluidic cell culture system. Anal Chem 79:8557–8563
6. Huang N-T, Chen W, Oh B-R et al (2012) An integrated microfluidic platform for in situ cellular cytokine secretion immunophenotyping. Lab Chip 12:4093–4101
7. Toriello NM, Douglas ES, Mathies RA (2005) Microfluidic device for electric field-driven single-cell capture and activation. Anal Chem 77:6935–6941
8. Mohammed JS, Wang Y, Harvat TA et al (2009) Microfluidic device for multimodal characterization of pancreatic islets. Lab Chip 9:97–106
9. Wu M-H, Huang S-B, Lee G-B (2010) Microfluidic cell culture systems for drug research. Lab Chip 10:939–956
10. Schultz NM, Huang L, Kennedy RT (1995) Capillary electrophoresis-based immunoassay to determine insulin content and insulin secretion from single islets of Langerhans. Anal Chem 67:924–929
11. Kulkarni R (2004) The islet β-cell. Int J Biochem Cell Biol 36:365–371
12. Kennedy RT, Kauri LM, Dahlgren GM, Jung S-K (2002) Metabolic oscillations in beta-cells. Diabetes 51(Suppl 1):S152–S161
13. Henquin JC, Ishiyama N, Nenquin M et al (2002) Signals and pools underlying biphasic insulin secretion. Diabetes 51(Suppl 1): S60–S67
14. Simon C, Brandenberger G (2002) Ultradian oscillations of insulin secretion in humans. Diabetes 51(Suppl 1):S258–S261
15. Peschke E, Peschke D (1998) Evidence for a circadian rhythm of insulin release from perifused rat pancreatic islets. Diabetologia 41:1085–1092

16. Shackman JG, Dahlgren GM, Peters JL, Kennedy RT (2005) Perfusion and chemical monitoring of living cells on a microfluidic chip. Lab Chip 5:56–63
17. Dishinger JF, Reid KR, Kennedy RT (2009) Quantitative monitoring of insulin secretion from microfluidic chip. Anal Chem 81: 3119–3127
18. Reid KR, Kennedy RT (2009) Continuous operation of microfabricated electrophoresis devices for 24 hours and application to chemical monitoring of living cells. Anal Chem 81:6837–6842
19. Han BG, Hao C-M, Tchekneva EE et al (2008) Markers of glycemic control in the mouse: comparisons of 6-hand overnightfasted blood glucoses to Hb A1c. Am J Physiol Endocrinol Metab 295:E981–E986
20. Gao N, Le Lay J, Qin W et al (2010) Foxa1 and Foxa2 maintain the metabolic and secretory features of the mature beta-cell. Mol Endocrinol 24:1594–1604
21. Jacobson SC, Ermakov SV, Ramsey JM (1999) Minimizing the number of voltage sources and fluid reservoirs for electrokinetic valving in microfluidic devices. Anal Chem 71: 3273–3276
22. Shackman JG, Watson CJ, Kennedy RT (2004) High-throughput automated post-processing of separation data. J Chromatogr A 1040: 273–282
23. Pralong WF, Bartley C, Wollheim CB (1990) Single islet beta-cell stimulation by nutrients: relationship between pyridine nucleotides, cytosolic Ca^{2+} and secretion. EMBO J 9:53–60

Chapter 6

On-Chip Magnetic Particle-Based Immunoassays Using Multilaminar Flow for Clinical Diagnostics

Mark D. Tarn and Nicole Pamme

Abstract

Magnetic particles have become popular in recent years for immunoassays due to their high surface-to-volume ratio and the ease of their manipulation. However, such assays also require multiple reaction and washing steps that are both time-consuming and manually laborious. Here, we describe a setup and methodology for performing rapid immunoassays on magnetic particles in continuous flow via their deflection through multiple laminar flow streams of reagents and washing solutions. In particular, we focus on the use of the microfluidic platform for a C-reactive protein (CRP) sandwich immunoassay in less than 60 s.

Key words Continuous flow, Deflection, Immunoassays, Microfluidics, Microparticles, Multilaminar flow

1 Introduction

The use of functionalized magnetic particles in clinical diagnostics has enabled vast reductions in immunoassay procedural times. This is largely due to the ability to rapidly exchange reagent and washing solutions while maintaining the particles inside the sample vial using a permanent magnet. Their small size (0.2–100 μm), high surface-to-volume ratio, ease of manipulation via a permanent magnet, and the vast range of surface functional groups (e.g., antigens, antibodies, DNA, chemical groups) makes them suitable for a number of biochemical applications. However, conventional magnetic particle-based assays require multiple reagent and washing steps that require laborious pipetting of solutions to-and-from a sample vial, and so the methodology is still relatively time-consuming.

The combination of microfluidic technology [1, 2] with magnetic particles brings several advantages over conventional magnetic particle-based procedures [3–6]. Microfluidic channels reduce the volumes of reagents required, waste production, device footprint, and diffusion times. The miniaturized dimensions of microfluidic devices and the channels within them allow permanent

Valérie Taly et al. (eds.), *Microchip Diagnostics: Methods and Protocols*, Methods in Molecular Biology, vol. 1547, DOI 10.1007/978-1-4939-6734-6_6, © Springer Science+Business Media LLC 2017

magnets to be placed near to the region of interest for particle manipulation, thus allowing large forces to be applied, while lessening the distance over which particles must be moved through the channels. Many magnetic particle-based microfluidic platforms involve a form of "trap-and-release" methodology, in which a plug of particles is trapped in a channel while reagents and washing buffer solutions are flushed consecutively over it. However, this retains the multi-step regime of the conventional technique, requiring the sequential exchange of solutions, and so suffers from the same drawback of being relatively time-consuming.

An elegant method of performing multi-step particle-based procedures is to perform them in continuous flow using a "multilaminar" flow platform [7]. Here, multiple laminar flow streams are typically generated across a chamber in the *x*-direction, with the flows featuring alternating reagents and washing buffer solutions. Particles are introduced into the chamber and while following the direction of flow they are also deflected in the *y*-direction, such that they sequentially cross each reagent and buffer stream, allowing consecutive reactions to take place on the particle surfaces with washing steps in-between. The multilaminar flow platform enables multi-step reactions and assays to take place in a single, extremely rapid process, and can be achieved by applying a number of forces on the particles to cause their deflection through the streams. Such forces have included micropillars/posts [8–10], microrail systems [11, 12], dielectrophoresis [13], acoustophoresis [14], pinched flow [15], hydrodynamic filtration [16, 17], optical forces [18], and biomolecular motors [19], while the use of inertial forces [20] and standing surface acoustic waves (SSAW) [21] has also been demonstrated.

Magnetic forces have seen considerable use in multilaminar flow platforms. This is easily achieved by generating laminar flow in a reaction chamber of alternating reagents and washing buffers via a syringe pump, and by placing a permanent magnet on the top of (or beside) the chamber to deflect the magnetic particles (*see* Fig. 1). Such a system provides a very simple setup that can perform particle-based processes in less than a minute compared to around an hour. We have applied this setup for several applications including immunoassays [22–25], DNA hybridization [23, 26], and the coating of magnetic templates [27]. Other groups have employed similar platforms for immunoassays [28–34], DNA extraction and washing [35, 36], and particle coating [37–39].

Here, we discuss the application of the multilaminar flow platform for performing sandwich immunoassays using C-reactive protein (CRP), a biomarker found in blood that increases with inflammation and infection, as an example [25], though the setup is largely applicable to other assays and reactions.

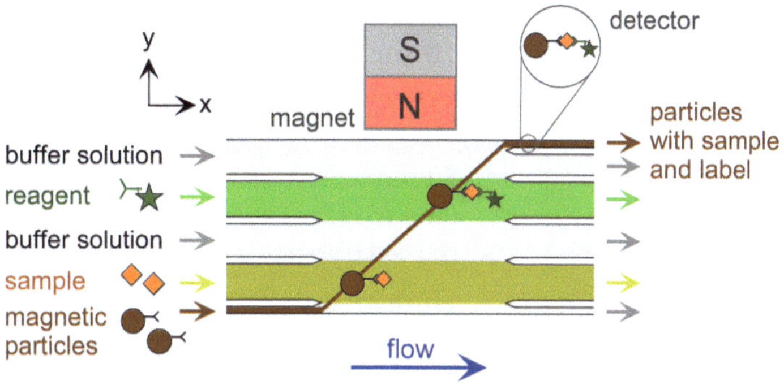

Fig. 1 Schematic demonstrating the multilaminar flow platform for a sandwich immunoassay. Magnetic particles functionalized with a primary antibody are continuously deflected through alternating laminar flow streams of reagents and washing solutions, allowing sequential reactions to take place on the particle surfaces

2 Materials

Prepare all buffer solutions and reagents using ultrapure water (18.2 MΩ cm at 25°C) and analytical grade reagents. Prepared buffer solutions should be filtered through a 0.22 μm filter prior to their use. Solutions can also be degassed by placing the bottles in an ultrasonic bath or via a vacuum to help avoid bubble formation in the microfluidic chip. Most assays performed using the methodology described here have been undertaken at room temperature (20°C), but the temperature requirements of a given assay should be considered.

2.1 Buffer Solutions, Biomolecules, and Magnetic Particles

1. Phosphate buffered saline (PBS) solution, pH 7.45 (*see* **Note 1**): Dissolve a 5 g PBS tablet (Gibco, UK) in about 400 mL water, add 0.05 g bovine serum albumin (BSA) added (to give a final 0.01 % w/v solution, *see* **Note 2**), and make up to 500 mL with water. Store at 4°C.
2. Tris reconstitution buffer (20 mM tris(hydroxymethyl)aminomethane (Tris), pH 8.0, 0.1 % (w/v) BSA): Dissolve 2.42 g Tris and 1 g BSA in about 900 mL water in a beaker (*see* **Note 3**). Add hydrochloric acid (*see* **Note 4**) to adjust the pH to 8.0. Make up to 1 L with water and store at 4°C.
3. C-reactive protein (CRP): Lyophilized recombinant human C-reactive protein (R&D Systems, UK), reconstituted in 1 mL Tris reconstitution buffer to a concentration of 200 μg mL^{-1} (*see* **Note 5**).
4. Primary CRP antibody (1° Ab): Lyophilized biotinylated monoclonal mouse antihuman C-reactive protein (R&D Systems, UK), reconstituted in 1 mL PBS solution to a concentration of 50 μg mL^{-1} (*see* **Note 6**).

5. Fluorescently labeled secondary CRP antibody (2°Ab-FITC): 10 mg mL^{-1} aqueous solution of polyclonal goat antihuman C-reactive protein conjugated to fluorescein isothiocyanate (λ_{ex} = 495 nm, λ_{em} = 521 nm) (Abcam plc., UK).
6. Superparamagnetic particles: Dynabeads M-270 Streptavidin (Invitrogen, UK), 2.8 μm diameter, streptavidin-coated, 6.5 × 10^8 particles mL^{-1} stock suspension (*see* **Note 7**).
7. Human serum samples: Store serum samples at −20°C in polypropylene tubes, and thaw immediately prior to use.

2.2 Microfluidic Chip Design and Fabrication

1. In a suitable photomask design software (e.g., AutoCAD, SolidWorks, L-Edit, Cadence), reproduce the chip design shown in Fig. 2. The mask design features five inlet and five outlet channels (240 μm width), each of which branches into two narrower channels (120 μm width) that connect to an 8 mm long × 3 mm wide reaction chamber (*see* **Notes 8–10**).
2. Fabricate the chip design in 1 mm thick glass (e.g., B-270 glass coated with chromium and photoresist layers, from Telic Company, CA, USA) (*see* **Note 11**) to a depth of 20 μm using conventional photolithography and wet etching procedures [40].
3. Drill access holes (400 μm diameter) using a carbide drill (Drill Service Ltd., UK).

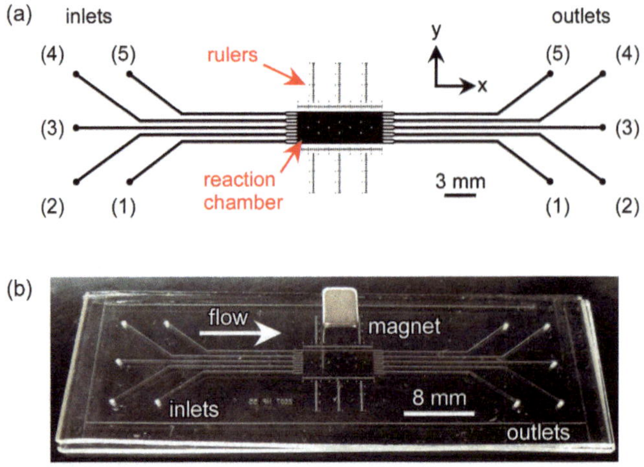

Fig. 2 (**a**) CAD design of the microfluidic device, featuring five inlets and five outlets, a reaction chamber (8 × 3 mm), and rulers for assisting magnet positioning. (**b**) Photograph of the final device fabricated in glass, with channels etched to a depth of 20 μm, and a NdFeB magnet positioned on the top of the chip, beside the chamber. Reprinted with permission from [25]. Copyright 2014 American Chemical Society

4. Remove the chromium and photoresist layers, wash the etched plate and a 1 mm thick glass top plate, and thermally bond the two plates together in a furnace.

2.3 Microfluidic Chip Holder Setup

1. Fabricate from aluminum (e.g., via a milling machine). A lower plate features a viewing window over which the reaction chamber of the chip can be placed. Two top plates are screwed to the bottom plate on each side of the viewing window. The top plates clamp the microfluidic chip in place, and feature threaded (6–40 UNF thread) access holes in a 5 × 5 mm matrix that are aligned with the chip's inlet and outlet holes).
2. Connectors for chip holder (IDEX Health & Scientific): (1) SuperFlangeless TinyTight 6–40 flat bottom PEEK nut for 1/16 inch OD tubing (part # M-644-03x), (2) 6–32 flat bottom ferrule for 360 μm tubing (part # N-123-03x).
3. Syringe adapters with Luer fittings (IDEX Health & Scientific): (1) Female Luer to ¼ inch-28 flat bottom female PEEK syringe adapter (part # P-200R), (2) ¼ inch-28 flat bottom Tefzel flangeless ferrule for 1/16 inch OD tubing (part # P-658), (3) ¼ inch-28 flat bottom PEEK flangeless nut for 1/16 OD tubing (part # P-235x).
4. PTFE tubing: 0.3 mm ID, 1.58 mm OD (Supelco, UK).
5. Fused silica capillary: 150 μm ID, 363 μm OD, polyimide coated (CM Scientific, UK).
6. Tygon microbore tubing: 0.010″ ID, 0.030″ OD (Cole-Parmer, UK).

2.4 On-Chip Assay Setup

1. Neodymium-iron-boron (NdFeB) magnet (Magnet Sales, UK) (*see* **Note 12**).
2. Blu-Tack (Bostik).
3. Ethanol.
4. Disposable syringes (e.g., 1 mL BD Plastipak).
5. Syringes (e.g., 1 mL SGE Gas Tight glass syringes with a fixed Luer lock).
6. Waste collection vials (e.g., 1.5 mL polypropylene vials, or a small glass vials).
7. Syringe pump capable of holding at least five syringes (e.g., a Harvard Apparatus PHD 22/2000 Infuse/Withdraw syringe pump with a 6/10 Multi-Syringe Rack attachment).
8. Inverted fluorescence microscope (e.g., a Nikon Eclipse Ti with a B-2A filter cube) with a suitable camera (e.g., a Retiga-EXL cooled charged couple device (CCD) camera from Media

Cybernetics, UK) and image/video acquisition software (e.g., Image-Pro Premier from Media Cybernetics, UK).

9. Image analysis software (e.g., ImageJ, available from http://imagej.nih.gov/ij/).

3 Methods

3.1 Activation of Magnetic Particles

1. Add 40 μL of 1° Ab solution (50 μg mL^{-1}) to a 1.5 mL polypropylene tube (e.g., an Eppendorf tube), followed by 160 μL of PBS solution to give a final 1° Ab concentration of 10 μg mL^{-1}. Vortex for 20 s.
2. Add 10 μL of stock Dynabeads M-270 Streptavidin suspension to the polypropylene tube containing the 1° Ab solution. Vortex for 20 s, and allow to incubate for 15 min to bind the biotinylated 1° antibodies to the streptavidin groups on the particles.
3. Performing the following washing step to the particles:
   (a) Place a permanent magnet next to the polypropylene tube for 2 min to allow the magnetic particles to be pulled to the side of the tube (*see* **Note 13**).
   (b) While the magnet is still in place, remove the supernatant using a pipette (*see* **Note 14**).
   (c) Add 1000 μL of PBS solution and vortex for 20 s.
   (d) Repeat **steps a–c** two more times.
4. Suspend the washed particles in 1000 μL of PBS solution to give a final particle concentration of 6.7×10^6 particles mL^{-1} (*see* **Note 15**). Vortex for 20 s.

3.2 Preparation of Solutions for On-Chip Assay

1. Dilute the CRP stock in 1000 μL to the desired concentration (e.g., 0.1–10 μg mL^{-1}) for the on-chip assay. Vortex to ensure thorough mixing.
2. Dilute the 2° Ab-FITC stock in 1000 μL to a concentration of 100 μg mL^{-1} for the on-chip assay. Vortex to ensure thorough mixing.

3.3 Preparation of Microfluidic Chip

1. Place the microfluidic chip onto the bottom plate of the aluminum chip holder, such that the reaction chamber is situated over the viewing window (*see* Fig. 3).
2. Screw the top plates of the chip holder onto the bottom plate, but do not fully tighten yet.
3. Carefully align the centers of the inlet and outlet holes of the chip with the connecting holes of the chip holder top plates, then tighten the top plates down (*see* **Note 16**).

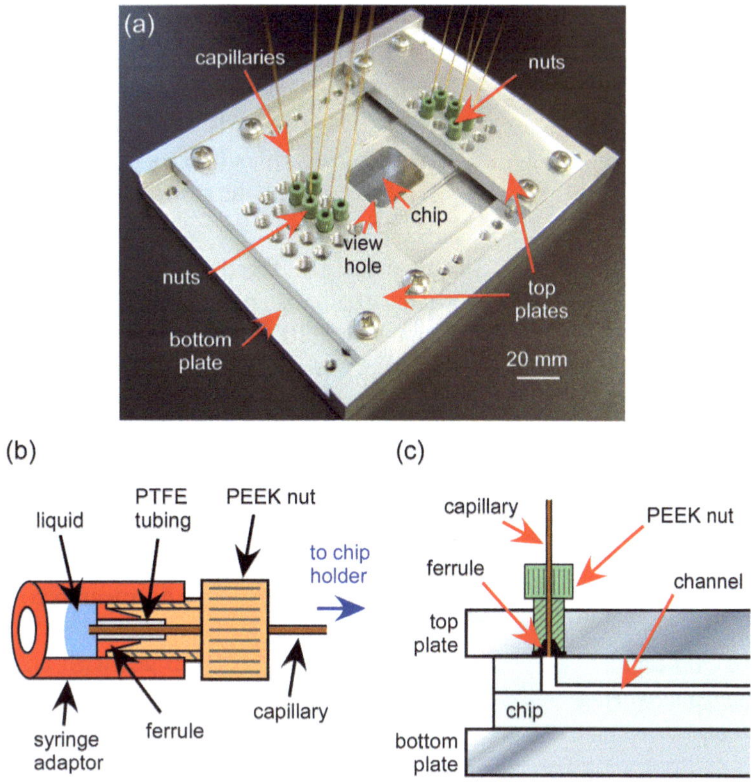

Fig. 3 (**a**) Photograph of the aluminum chip holder showing the connection of the inlet and outlet capillaries to the microfluidic chip. Reprinted with permission from [25]. Copyright 2014 American Chemical Society. (**b**) Schematic of the syringe adapter setup at the open end of the inlet capillaries. (**c**) Schematic showing the interfacing of the capillaries to the microfluidic device via the chip holder. Reprinted with permission from [25]. Copyright 2014 American Chemical Society

4. Using a ceramic cutting tile, cut five sections of fused silica capillary to size for use as inlets, such that they can reach from the chip to the syringe pump (e.g., 20 cm).
5. Cut five short sections of fused silica capillary to size for use as outlet tubing (e.g., 5 cm).
6. To one end of each piece of capillary, attach a ferrule (part # N-123-03x) (*see* **Note 17**) and place a PEEK nut (part # M-644-03x) over it.
7. Screw the inlet and outlet capillaries into the chip holder such that they are connected to the respective access holes on the chip.
8. To the open ends of the outlet capillaries, attach short sections of Tygon tubing and place the other in a waste collection vial (*see* **Note 18**).

9. Place PEEK nuts (part # P-235x) over the open ends of the inlet capillaries.
10. Cut short sections (e.g., 0.8 mm) of PTFE tubing and place these over the ends of the inlet capillaries (*see* **Note 19**), then place the ferrules (part # P-658) over the PTFE sections. The flat part of the ferrule should be flush with the edge of the PTFE sleeve.
11. Position the ferrule and its sleeve such that the open end of the capillary protrudes slightly from the ferrule/sleeve (*see* **Note 20**).
12. Pull the PEEK nut around the ferrule and screw the syringe adapter (part # P-200R) on tightly to seal the adapter setup. Repeat for the remaining inlet capillaries.
13. Place the chip holder onto the sample stage of the fluorescence microscope. Fix in place with tape or Blu-Tack if necessary.
14. Place the syringe pump next to the microscope, such that the inlet capillaries can be connected to syringes on the pump.

3.4 On-Chip Multilaminar Flow Assay

1. Fill five disposable syringes with ethanol, remove any air present, and place them on the syringe pump rack. Either engage the syringe pump manually or via the pump controls to pressurize the syringes to the point that solution is present at the tips of the syringes (i.e., there is no air present in the syringe).
2. Attach the inlet capillaries of the chip holder to the syringes and flush the chip at 100 $\mu L\ h^{-1}$ for at least 5 min.
3. Stop the pump and remove the syringes from it, but do not disconnect the syringes from the capillaries yet.
4. Fill five syringes with purified water and place them on the pump, and adjust the pump until the syringes are under pressure with liquid present at the tips of the syringes.
5. Carefully remove the syringes of ethanol from the syringe adapters while leaving a small amount of liquid in the adapter to prevent air from entering the capillary (*see* **Note 21**).
6. Partially fill the syringe adapter with water, which helps to prevent air bubbles entering the chip once the adapters are connected to the syringes (*see* **Note 22**).
7. Carefully connect the syringe adapters to the water-filled syringes on the pump, and flush the chip with water at 100 $\mu L\ h^{-1}$ for at least 5 min.
8. Repeat **steps 3–7** with PBS buffer, and flush the chip at 100 $\mu L\ h^{-1}$ for at least 5 min.
9. Fill glass syringes with (1) activated magnetic particle suspension (vortexed immediately beforehand), (2) CRP standard solution or serum sample, (3) 2° Ab-FITC solution, and (4) two syringes of PBS solution.

10. Repeat **steps 3–7** again, but this time place the filled glass syringes onto the syringe pump. The magnetic particle suspension should be attached to the capillary for inlet 1, the CRP standard/sample to inlet 2, the 2° Ab-FITC to inlet 4, and the PBS syringes to inlets 3 and 5 (*see* **Note 23**).
11. Start pumping at 100 μL h^{-1} for several seconds before lowering the flow rate to 10 μL h^{-1}, and be aware of air bubbles that may have been introduced into the chip (*see* **Note 24**).
12. Once the laminar flow has stabilized (*see* **Note 25**) and the magnetic particles are flowing through the reaction chamber (*see* **Note 26**), place the NdFeB magnet onto the chip and position it such that the magnetic particles are deflected from the channel of inlet 1 to the channel of outlet 5. Fix the magnet in place, for example using tape, Blu-Tack, or a bespoke holder (*see* **Note 27**).
13. Capture images/videos of the moving particles passing through the chamber both before and after passing through the fluorescent 2° Ab-FITC laminar flow stream, and analyze the fluorescence intensity/fluorescence peak area (*see* **Note 28**) of the particles in the images (*see* Fig. 4).
14. Replace the CRP standard/serum sample syringe (connected to inlet 2) as necessary with different standards or samples to generate a calibration graph and determine biomarker concentrations in samples. It is recommended to first flush that capillary with a syringe of PBS solution for several minutes to avoid contamination, before replacing it with the new CRP standard/serum sample syringe.

4 Notes

1. While PBS is an appropriate running buffer here in terms of keeping the extent of particle agglomeration low, other biomolecules (e.g., antibodies, antigens) having different surface properties may require different buffer solutions to maintain stable suspensions. The surface charges of the biomolecules and the magnetic particles at each stage of the assay should be considered at the buffer pH used, and in relation to the surface charge of the chip surface. For example, when using glass chips, it is often preferential to use a buffer pH > 4 to maintain a negative surface charge. In this case, the buffer pH should ideally be maintained at a value greater than the isoelectric point (pI) of the biomolecules in order that they, and the magnetic particles on which they are bound, exhibit negatively charged surfaces to repel them from each other and from the glass surfaces.

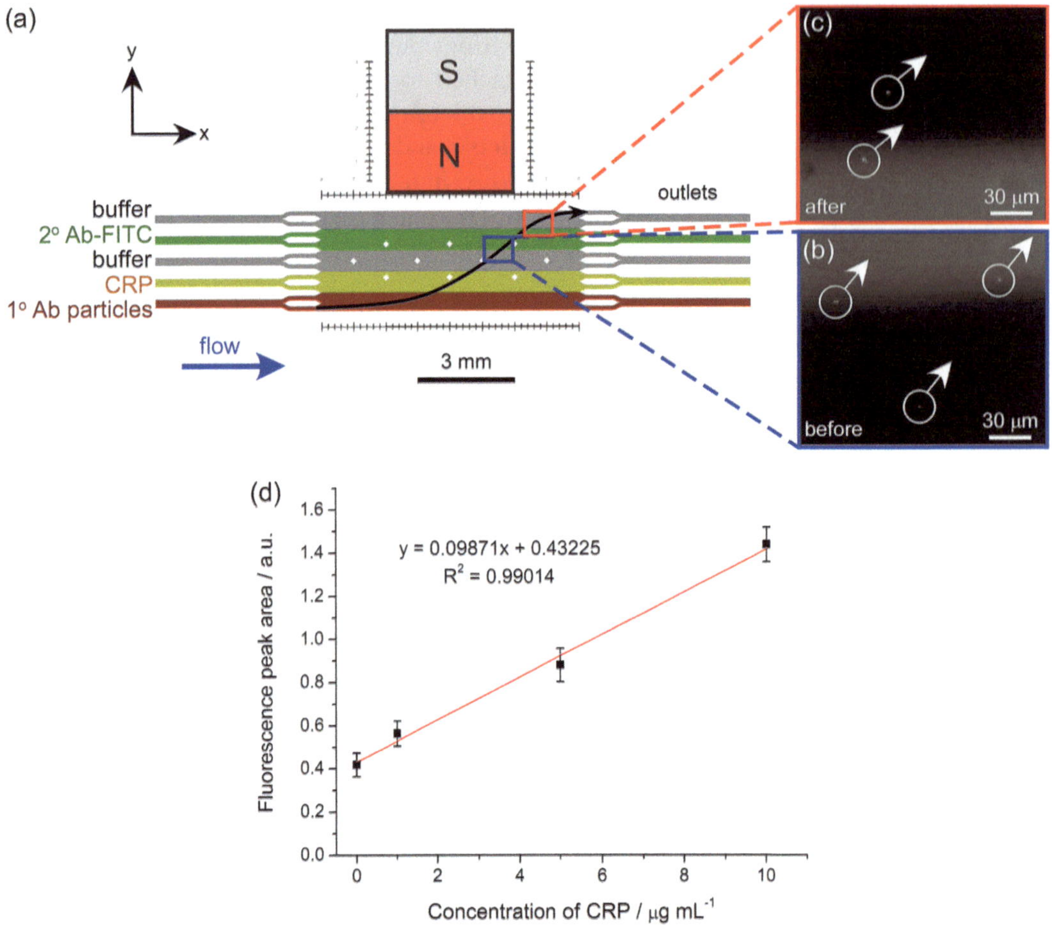

Fig. 4 (**a**) Schematic of the C-reactive protein (CRP) setup with photographs showing (**b**) magnetic particles as they are entering the fluorescent 2°Ab-FITC stream, and (**c**) the now fluorescent particles as they leave the 2°Ab-FITC stream. (**d**) Calibration graph of the fluorescence signal of CRP standards. Reprinted with permission from [25]. Copyright 2014 American Chemical Society

2. BSA is added to the PBS solution to reduce the sticking of magnetic particles to each other and to the walls of the microfluidic channel.
3. Having some water already present in the beaker, with a magnetic stirrer bar and a stirrer hot plate, helps to dissolve the Tris relatively quickly.
4. Decreasing concentrations of HCl can be used to bring the pH closer to 8.0. For example, 1 M HCl can first be used to bring the pH close to 8, before using lower concentrations (e.g., 0.1 M HCl, 0.01 M HCl) for final adjustments to pH 8.0. Care should be taken that the pH does not drop below pH 8.0, hence the use of lower HCl concentrations when near to the endpoint.

5. Reconstituted CRP can be stored in a fridge (2–8°C) for one month, or can be aliquoted and stored in a freezer (−20 to −70°C) for three months. Repeated freeze-thaw cycles should be avoided.
6. Reconstituted primary CRP antibody (1° Ab) can be stored in a fridge (2–8°C) for one month, or can be aliquoted and stored in a freezer (−20 to−70°C) for six months. Repeated freeze-thaw cycles should be avoided.
7. While various other types of streptavidin-coated superparamagnetic particles can be employed, we have found that the Dynabeads M-270 range yields a strong attraction to the magnetic field [41].
8. To aid in the use of a chip holder, it is recommended to design the access holes of the chip to fit on a matrix (e.g., 5 × 5 mm hole-center-to-hole-center).
9. The addition of ruler structures to the microfluidic chip design, near to the chamber and in both the *x*- and *y*-directions, assists in the reproducible alignment of the permanent magnet with the chamber later.
10. Due to the width of the particle inlet channel, variation in the starting position of the particles (in the *y*-direction) in the chamber can lead to a spread of particle trajectories, which can result in variations in fluorescence signal due to the slightly different paths taken through the reagent streams. A flow focusing feature added to the chip design can allow the particles to be focused into a narrow stream prior to their entering the chamber, resulting in a more reproducible path through the chamber and smaller variations in fluorescence signal.
11. Polymer, rather than glass, microfluidic devices can also be used, and can be fabricated via techniques such as hot embossing or injection molding [42].
12. We have found that a 4 × 4 × 5 mm NdFeB magnet works well, but a variety of magnet sizes/types can be used. However, the operator must optimize the magnet size/type, its distance from the reaction chamber, and the flow speed of the magnetic particles through the reaction chamber (in the *x*-direction) to maintain a steady trajectory from one corner of the reaction chamber to the other.
13. Blu-Tack can be used to temporarily fix the tube and the magnet in place, while also allowing the magnet to be placed at the optimum position such that the particles are drawn against the side of the tube.
14. While particle loss can be expected during the removal of the supernatant, this can largely be reduced by having the particles pulled to the side of the tube, while the pipette tip used for

removing the solution is placed at the bottom of the tube, angled slightly away from the trapped particles.

15. Particle concentration is important, as too high a concentration can result in excessive agglomeration of particles.
16. Care must be taken not to crack the glass chip when screwing the top plates of the chip holder to the bottom plate. Small pieces of PTFE sheet can be cut to size and placed underneath the chip, below the locations of the top plates, which can help to cushion the chip and reduce the risk of cracking.
17. The end of the capillary should be flush with the flat of the ferrule, rather than protruding from the ferrule. This has the effect that the capillaries are not actually situated in the chip's access holes, but rather they are above the holes. This helps to stop air bubbles becoming trapped behind a section of capillary as is sometimes observed when the capillary is actually placed inside an access hole, which can lead to smaller air bubbles being continuously sheared off from the main bubble and entering the chamber.
18. The capillary from outlet 5 can be placed in its own collection vial if the operator is interested in collecting the deflected magnetic particles for further analysis. However, this vial should only be placed once the system has stabilized and particles are being deflected across the chamber.
19. It is recommended to push a separate piece of capillary (363 μm OD) through the PTFE tubing (300 μm ID) to bore out the PTFE to the correct size prior to its use with the capillaries in the chip holder. The capillary essentially "shaves off" a sliver of PTFE as it is pushed through, which can potentially block the capillary; hence, it is best to bore out the PTFE before using it on the experimental setup.
20. Having the capillary protruding slightly from the ferrule and its PTFE sleeve makes it easier for the particles to enter the capillary than if they were flush with each other, as the particles are seemingly repelled slightly from the PTFE.
21. Carefully twist the syringe while removing the syringe from the adapter and applying light pressure to the plunger. Twisting assists in the smooth removal of the syringe, while applying light pressure ensures that a small amount of liquid is left covering the capillary inlet inside the adapter. This assists in preventing air from entering the microfluidic device.
22. Place the syringe tip against the inside wall of the syringe adapter and slowly drop water down the side of the adapter. This helps to prevent an air bubble from being trapped in the adapter.

23. "Topping up" the syringe adapters with solution may not be necessary beforehand provided there is enough PBS buffer in the adapter to form a connection between the adapters and the syringes without introducing air.
24. If air bubbles enter the chamber and become trapped on the channel surfaces, they can usually be removed via a burst of pressure applied from the syringe pump, usually by simply increasing the flow rate and dropping it back down to 10 μL h^{-1}. In extreme cases by manually pushing the back plate on the syringe pump forward before allowing the laminar flow to stabilize at 10 μL h^{-1}.
25. The status and stability of the multilaminar flow regime can be monitored by observing the flow of the fluorescent 2° Ab-FITC stream.
26. Due to the density of the magnetic particles (1.6 g cm^{-3}) relative to the buffer solution (~1 g cm^{-3}), the particles are prone to sedimentation. Sedimentation is rarely a problem once the particles have entered the chip itself, but can be an issue in the syringe containing the magnetic particle suspension. To avoid these effects for a time, vortex the particle suspension prior to drawing it into a syringe. It can also help to carefully rotate the syringe in the pump from time to time, and allowing the laminar flow streams to stabilize again afterward.
27. If this experiment has been performed previously and the magnet position has been optimized, then the magnet can be positioned earlier on during the experimental setup.
28. Analyzing the fluorescence peak area of a moving magnetic particle, rather than only the fluorescence intensity, helps to account for variations in the depth of the particles in the chamber and the different trajectories of the particles across the chamber.

References

1. Tarn MD, Pamme N (2013) Microfluidics. In: Reedijk J (ed) Elsevier reference module in chemistry, molecular sciences and chemical engineering. Elsevier, Waltham, MA. doi:10.1016/B978-0-12-409547-2.05351-8
2. Nge PN, Rogers CI, Woolley AT (2013) Advances in microfluidic materials, functions, integration, and applications. Chem Rev 113(4).2550–2583
3. Pamme N (2006) Magnetism and microfluidics. Lab Chip 6(1):24–38
4. Pamme N (2012) On-chip bioanalysis with magnetic particles. Curr Opin Chem Biol 16(3–4):436–443
5. Gijs MAM (2004) Magnetic bead handling on-chip: new opportunities for analytical applications. Microfluid Nanofluid 1(1):22–40
6. Gijs MAM, Lacharme F, Lehmann U (2010) Microfluidic applications of magnetic particles for biological analysis and catalysis. Chem Rev 110(3):1518–1563
7. Tarn MD, Lopez-Martinez MJ, Pamme N (2014) On-chip processing of particles and cells via multilaminar flow streams. Anal Bioanal Chem 406(1):139–161
8. Morton KJ, Loutherback K, Inglis DW, Tsui OK, Sturm JC, Chou SY, Austin RH (2008) Crossing microfluidic streamlines to lyse, label and wash cells. Lab Chip 8(9):1448–1453

9. Kantak C, Beyer S, Yobas L, Bansal T, Trau D (2011) A 'microfluidic pinball' for on-chip generation of Layer-by-Layer polyelectrolyte microcapsules. Lab Chip 11(6):1030–1035
10. Sochol RD, Li S, Lee LP, Lin L (2012) Continuous flow multi-stage microfluidic reactors via hydrodynamic microparticle railing. Lab Chip 12(20):4168–4177
11. Chung SE, Park W, Shin S, Lee SA, Kwon S (2008) Guided and fluidic self-assembly of microstructures using railed microfluidic channels. Nat Mater 7(7):581–587
12. Chung SE, Park W, Shin S, Lee SA, Kwon S (2008) Guided fluidic self-assembly of microtrains using railed microfluidics. Paper presented at the FNANO08—5th Annual Conference on Foundations of Nanoscience, Snowbird Cliff Lodge, Snowbird, Utah, USA, 22–25 April 2008.
13. Seger U, Gawad S, Johann R, Bertsch A, Renaud P (2004) Cell immersion and cell dipping in microfluidic devices. Lab Chip 4(2):148–151
14. Augustsson P, Åberg L, Swärd-Nilsson A-M, Laurell T (2009) Buffer medium exchange in continuous cell and particle streams using ultrasonic standing wave focusing. Microchim Acta 164(3–4):269–277
15. Chiang Y-Y, West J (2013) Ultrafast cell switching for recording cell surface transitions: new insights into epidermal growth factor receptor signalling. Lab Chip 13(6): 1031–1034
16. Yang S, Ji B, Ündar A, Zahn JD (2006) Microfluidic devices for continuous blood plasma separation and analysis during pediatric cardiopulmonary bypass procedures. ASAIO J 52(6):698–704
17. Toyama K, Yamada M, Seki M (2012) Isolation of cell nuclei in microchannels by short-term chemical treatment via two-step carrier medium exchange. Biomed Microdevices 14(4):751–757
18. Eriksson E, Enger J, Nordlander B, Erjavec N, Ramser K, Goksor M, Hohmann S, Nystrom T, Hanstorp D (2007) A microfluidic system in combination with optical tweezers for analyzing rapid and reversible cytological alterations in single cells upon environmental changes. Lab Chip 7(1):71–76
19. Kim T, Cheng L-J, Kao M-T, Hasselbrink EF, Guo L, Meyhofer E (2009) Biomolecular motor-driven molecular sorter. Lab Chip 9(9):1282–1285
20. Tan AP, Dudani JS, Arshi A, Lee RJ, Tse HTK, Gossett DR, Di Carlo D (2014) Continuous-flow cytomorphological staining and analysis. Lab Chip 14(3):522–531
21. Li S, Ding X, Mao Z, Chen Y, Nama N, Guo F, Li P, Wang L, Cameron CE, Huang TJ (2015) Standing surface acoustic wave (SSAW)-based cell washing. Lab Chip 15(1): 331–338
22. Peyman SA, Iles A, Pamme N (2008) Rapid on-chip multi-step (bio)chemical procedures in continuous flow—manoeuvring particles through co-laminar reagent streams. Chem Commun 10:1220–1222
23. Peyman SA, Patel H, Belli N, Iles A, Pamme N (2009) A microfluidic system for performing fast, sequential biochemical procedures on the surface of mobile magnetic particles in continuous flow. Magnetohydrodynamics 45(3): 361–370
24. Peyman SA, Iles A, Pamme N (2009) Mobile magnetic particles as solid-supports for rapid surface-based bioanalysis in continuous flow. Lab Chip 9(21):3110–3117
25. Phurimsak C, Tarn MD, Peyman SA, Greenman J, Pamme N (2014) On-chip determination of C-reactive protein using magnetic particles in continuous flow. Anal Chem 86(21):10552–10559
26. Vojtíšek M, Iles A, Pamme N (2010) Rapid, multistep on-chip DNA hybridisation in continuous flow on magnetic particles. Biosens Bioelectron 25(9):2172–2176
27. Tarn MD, Fakhrullin RF, Paunov VN, Pamme N (2013) Microfluidic device for the rapid coating of magnetic cells with polyelectrolytes. Mater Lett 95:182–185
28. Sasso LA, Undar A, Zahn JD (2010) Autonomous magnetically actuated continuous flow microimmunofluorocytometry assay. Microfluid Nanofluid 9(2–3):253–265
29. Sasso L, Johnston I, Zheng M, Gupte R, Ündar A, Zahn J (2012) Automated microfluidic processing platform for multiplexed magnetic bead immunoassays. Microfluid Nanofluid 13(4):603–612
30. Sasso LA, Aran K, Guan Y, Ündar A, Zahn JD (2013) Continuous monitoring of inflammation biomarkers during simulated cardiopulmonary bypass using a microfluidic immunoassay device—A pilot study. Artif Organs 37(1):E9–E17
31. Zhou Y, Wang Y, Lin Q (2010) A microfluidic device for continuous-flow magnetically controlled capture and isolation of microparticles. J Microelectromech Syst 19(4):743–751
32. Lee SHS, Hatton TA, Khan SA (2011) Microfluidic continuous magnetophoretic

protein separation using nanoparticle aggregates. Microfluid Nanofluid 11(4):429–438
33. Gao Y, Lam AWY, Chan WCW (2013) Automating quantum dot barcode assays using microfluidics and magnetism for the development of a point-of-care device. ACS Appl Mater Interfaces 5(8):2853–2860
34. Ganguly R, Hahn T, Hardt S (2010) Magnetophoretic mixing for in situ immunochemical binding on magnetic beads in a microfluidic channel. Microfluid Nanofluid 8(6):739–753
35. Karle M, Miwa J, Czilwik G, Auwaerter V, Roth G, Zengerle R, von Stetten F (2010) Continuous microfluidic DNA extraction using phase-transfer magnetophoresis. Lab Chip 10(23):3284–3290
36. Karle M, Woehrle J, Miwa J, Paust N, Roth G, Zengerle R, von Stetten F (2011) Controlled counter-flow motion of magnetic bead chains rolling along microchannels. Microfluid Nanofluid 10(4):935–939
37. Tsai SSH, Wexler JS, Wan J, Stone HA (2011) Conformal coating of particles in microchannels by magnetic forcing. Appl Phys Lett 99(15):153509
38. Tsai SSH, Wexler JS, Wan J, Stone HA (2013) Microfluidic ultralow interfacial tensiometry with magnetic particles. Lab Chip 13(1): 119–125
39. Moon B-U, Hakimi N, Hwang DK, Tsai SSH (2014) Microfluidic conformal coating of non-spherical magnetic particles. Biomicrofluidics 8(5):052103
40. McCreedy T (2000) Fabrication techniques and materials commonly used for the production of microreactors and micro total analytical systems. TrAC, Trends Anal Chem 19(6): 396–401
41. Tarn MD, Peyman SA, Robert D, Iles A, Wilhelm C, Pamme N (2009) The importance of particle type selection and temperature control for on-chip free-flow magnetophoresis. J Magn Magn Mater 321(24):4115–4122
42. Becker H, Gärtner C (2008) Polymer microfabrication technologies for microfluidic systems. Anal Bioanal Chem 390(1):89–111

Chapter 7

Digital Microfluidics Assisted Sealing of Individual Magnetic Particles in Femtoliter-Sized Reaction Wells for Single-Molecule Detection

Deborah Decrop, Elena Pérez Ruiz, Phalguni Tewari Kumar, Lisa Tripodi, Tadej Kokalj, and Jeroen Lammertyn

Abstract

Digital microfluidics has emerged in the last years as a promising liquid handling technology for a variety of applications. Here, we describe in detail how to build up an electrowetting-on-dielectric-based digital microfluidic chip with unique advantages for performing single-molecule detection. We illustrate how superparamagnetic particles can be printed with very high loading efficiency (over 98 %) and single-particle resolution in the microwell array patterned in the Teflon-AF® surface of the grounding plate of the chip. Finally, the potential of the device for its application to single-molecule detection is demonstrated by the ultrasensitive detection of the biotinylated enzyme β-Galactosidase captured on streptavidin-coated particles in the described platform.

Key words Digital Microfluidics, Microwell Array, Magnetic Particles, Single-Molecule Detection, Digital Bioassays

1 Introduction

Bioassay development has been constantly stimulated for achieving a greater analytical sensitivity, not only for pushing the limits of detection of lower target concentrations, but also for improving the precision, minimizing the interferences, and enabling smaller sample volumes for analysis. In this context, the highest sensitivity that can be aimed is the single-molecule resolution and, given the recent developments in the field of single-molecule detection, we believe that this paradigm holds great potential in the future of bioassays. This proves to be extremely important for early-stage diagnosis of many diseases and therefore, imperative for improving the existing diagnostic tools. Optical single-molecule detection methods find their origin in the work of Noji et al. [1] and Walt et al. [2, 3]. They demonstrated the isolation of individual target biomolecules in ultra-small reaction chambers, where individual

Valérie Taly et al. (eds.), *Microchip Diagnostics: Methods and Protocols*, Methods in Molecular Biology, vol. 1547,
DOI 10.1007/978-1-4939-6734-6_7,

fluorescent signals were generated and detected. By encapsulating enzymes in small volumes (~50 fL) containing the fluorogenic substrate, the enzymatic product was prevented from diffusing. Therefore, only a single enzyme molecule was sufficient for the generation of a detectable signal. In this way, quantification of individual molecules was performed by counting the number of fluorescent wells. Recently, this concept has been further optimized by the implementation of magnetic particles for capturing individual molecules and then confining them into the femtoliter-sized reaction wells [4–8], making digital assays simpler and more efficient. In these magnetic particle-based digital assays, the biomolecules are first captured on the particles surface. Because the target concentration in the sample is very low, the ratio of target molecules to particles is consequently very small. In this case, the percentage of particles that capture a target molecule follows a Poisson distribution, which predicts that the magnetic particles will carry either one single target molecule or none. Rissin et al. [8] were one of the first to apply magnetic particles in the context of digital assays. In their platform, microwell arrays were fabricated on an optical fiber bundle and magnetic particles were loaded in the microwells with the assistance of centrifugal forces. Subsequently, the particles were sealed with a silicone gasket in the presence of a droplet containing the fluorogenic substrate. Significantly lower limits of detection were attained on their platform, as compared with conventional ELISA techniques [4]. However, the centrifugal-based loading process leads to an arbitrary distribution of the particles in the microwell array and the achieved loading efficiency was only about 60 %. This implied that for a digital quantification the wells containing a particle required previous identification for precisely determining the ratio of "on" and "off" wells. Due to this shortcoming, the throughput and efficiency of the system suffered.

We have recently introduced electrowetting-on-dielectric-based digital microfluidics (EWOD-DMF) as a promising tool for seeding and sealing single magnetic particles into femtoliter-sized reaction wells [9]. EWOD-DMF is based on the concept of software-assisted manipulation of discrete, independently addressable liquid droplets, upon the application of electric potentials that alter the contact angle between the droplet and the surface and therefore, the surface wettability [10–12]. During the last decade, DMF has emerged as a promising liquid handling technology for a variety of applications [13–19], as several demonstrations have validated its potential in terms of miniaturization and automation and, in addition, it is highly reconfigurable and reprogrammable. Moreover, in contrast with other microfluidics techniques (i.e., channel-based), DMF offers the exclusive advantage for precisely handling solid microsized items, such as magnetic particles [20]. In the presented platform, microarrays containing more than

60,000 hydrophilic-in-hydrophobic (HIH), femtoliter-sized wells are patterned on the grounding plate of a DMF chip, while actuation electrodes are fabricated in the actuation plate using standard lithographic techniques. For manipulating droplet using EWOD actuation on the DMF system, an electronic interface was developed that translates user-defined commands into voltage actuation. After the DMF plate assembly, a permanent magnet is placed above the microwell array and the droplet containing magnetic particles is transported over the microwell array. This leads to the capturing of particles in the spatially defined microwells with single particle resolution. With this approach, particle loading efficiencies of nearly 100 % can be achieved. This potentially avoids the need of manually counting the wells that contain a magnetic particle. In addition, since DMF offers the possibility of integrating all fluid manipulations on-chip, fully automated single-molecule assays can be performed on this platform. As a proof of concept, we quantified the biotinylated protein β-Galactosidase down to attomolar levels in the developed DMF chip. In addition, the potential of the herein-reported DMF platform is currently being explored for more complex single-molecule assays for the detection of protein, DNA molecules, and cells. A flow chart summarizing the outline of the paper is shown in Fig. 1.

2 Materials

1. Software: AutoCAD 2013, student version, USA.
2. Unpolished Borofloat® 33 4′ wafers, Schott, Germany.
3. Acetone and isopropanol, Sigma-Aldrich, USA.
4. Microposit S1818 positive photoresist, Rohm and Haas, USA.
5. Microposit 351-developer, Rohm and Haas, USA.
6. Cyantec CR-7 chromium etchant, KMG electronic chemicals, USA.
7. Silane A174, Sigma-Aldrich, USA.
8. Parylene coating services, Rosenheim, Germany.
9. Teflon-AF® 1600, DuPont, USA.
10. Perfluoro-compound FC-40®, Sigma-Aldrich, USA.
11. Conductive Copper Foil Tape, 3 M™, Belgium.

Fig. 1 Overview of the structure of this protocol

12. Cover glasses, 0.7 mm thickness, Menzel-Gläser, Braunschweig, Germany.
13. Fluoralkylsilane Dynasylan F8263, Evonik, Germany.
14. Aluminum etch: Prepare a stock of 1 L that you can store in a glass bottle. Mix 800 mL H_3PO_4 with 50 mL HNO_3, 100 mL deionized water, 50 mL CH_3COOH, and 200 μL of surfactant (NCW 1001).
15. PCB board.
16. N-channel MOSFET, IRFD210, International Rectifier, USA.
17. I/O hardware device, PCI-6509 and compatible cables, National Instruments Corp, USA.
18. Software: LabView2014,National Instruments Corp., USA.
19. Software: Txt generator MATLAB R2014b, TheMathWorks, Inc., USA.
20. Board to board connectors and compatible cables.
21. Frequency generator, GFG-8216 A-ISO-TECH, England.
22. Amplifier, FLC Electronics A600, Origin Sweden.
23. Red and black connecting wires.
24. Silver conductive paint, Electrolube, UK.
25. Buffer samples, e.g., PBS, cell medium, human serum, etc.
26. Vapor-Lock™ oil, Qiagen, Hilden Germany.
27. Vacuum pen, Ted Pella Inc., USA.
28. Lodestar 2.7 Streptavidin superparamagnetic particles, 10.5 mg/mL, 2.7 μm in diameter, Agilent Technologies, Belgium.
29. Magnet, Ø 4 mm, 4 mm high, Webcraft GmbH, Germany.
30. Biotinylated β-Galactosidase (BβG), Thermo Scientific, USA.
31. Bovine Serum Albumine (BSA), Sigma-Aldrich, USA.
32. Tween 20, Sigma-Aldrich, USA.
33. Pluronic F-68, Sigma-Aldrich, USA.
34. Phosphate buffered Saline (PBS buffer): 137 mM NaCl, 2.7 mM KCl,10 mM Na_2HPO_4, 1.8 mM KH_2PO_4 dissolved in deionized water.
35. Blocking Buffer: PBS containing 1 % BSA and 0.05 % Tween-20.
36. Washing Buffer: PBS with 0.1 % BSA and 0.1 % Tween-20.
37. EWOD-manipulation Buffer: PBS with 0.05 % Pluronic F-68.
38. FDG substrate (β-D-galactopyranoside), Sigma Aldrich, USA.
39. Surface Profiler, e.a. Dektak 3030 Surface Profiler.
40. Magnetron sputtering Source (Balzers BAE 370, Pfäffikon, Switzerland).

3 Methods

3.1 Photomasks

1. Design an actuation electrode layout, in AutoCAD software, with actuation electrodes of 2.8 × 2.8 mm^2, each with an inter-electrode gap of 40 μm. All electrodes should be connected to an individual electrode pad that electrically connects the electrodes with the chip holder. Figure 2 illustrates an electrode layout, designed in AutoCAD software, to fabricate 4 chips simultaneously (*see* **Note 1**).
2. Design a 2 × 2 mm^2 microwell array layout, in AutoCAD software, consisting of a hexagonal array of 62,500 circles with a diameter of 2.8 μm and a center-to-center distance of 8 μm.
3. Fabricate the masks on a chrome-on-glass photomask using a LaserPattern generator. Alternatively, the mask generation can be outsourced (*see* **Note 2**).

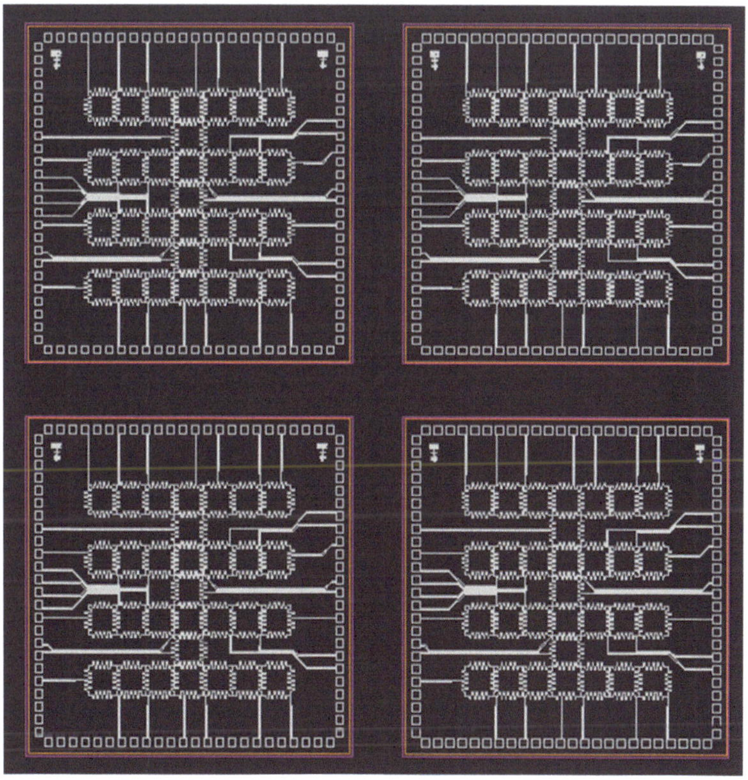

Fig. 2 Tree-structure electrode layout, designed using AutoCAD software. *Outer squared lines* are dicing marker lines, alignment markers in *upper left* and *right corner*, contact pads connected to electrodes with thin wires (100 μm wires that narrow down to 30 μm in the proximity of the electrodes)

3.2 Fabrication of Actuation Plate

1. Clean 4 inch glass wafers by immersing them respectively 5 min in acetone, isopropanol, and deionized water. Repeat this procedure three times and blow dry the wafers with a nitrogen gun.
2. Deposit a 100 nm thin layer of chromium using magnetron sputtering. Create a glow discharge plasma with a low radio frequency (RF) at 150 W when a chamber pressure of $2x10^{-3}$ mbar is reached. First pre-sputter a blank wafer for 10 min and then sputter the glass wafers for 2.5 min resulting in a 100 nm thick chromium layer. The layer thickness can be verified using a Surface Profiler (*see* **Note 3**).
3. Rinse the chromium sputtered glass wafers twice with water and dry them under nitrogen. Spincoat the cleaned wafers with S1818 positive photoresist for 10 s at 3000 rpm (~510 × *g*) and bake for 9 min at 90 °C on a preheated hot plate.
4. Place the wafer on an UV-pattern generator and align the actuation electrode layout photomask, bearing the electrode layout. Bring the wafer and photomask in close contact and UV-expose the photoresist 40 s. at 5 mW/cm^2. Develop the photoresist for 1 min in 351-developer (4× diluted in deionized water), and rinse with deionized water.
5. Wet etch the underlying chromium layer by submerging the wafers in Cyantec CR-7 chromium etchant at room temperature for 60 s, until all exposed chromium regions are removed (uncovered by photoresist). After wet etching, wash the wafers in an automated water bath and dry the wafers using the nitrogen gun.
6. Dice each glass wafer into four individual plates showing the electrode layout. The photomask contains marker lines to facilitate the alignment in this step. After dicing, check that each plate fits in the chip holder (*see* Subheading 3.4) (*see* **Note 4**).
7. Strip the remaining photoresist by submerging the wafers for 15 min in a preheated acetone bath at 50 °C. Afterward, rinse the plates with deionized water and dry with a nitrogen gun.
8. Clean the plates with O_2 plasma reactive ion etching (150 mTorr, 100 W, 32 sccm O_2) for 15 min and subsequently coat them with a 3.5 μm thick Parylene-C layer using chemical vapor deposition (see materials). Prior to Parylene-C coating, prime the plates with silane A174, to improve the adhesion of Parylene-C on the chromium electrodes (*see* **Note 5**).
9. Dilute Teflon-AF® (6 % w/w) to a 3 % w/w in Fluorinert FC-40. Immediately after Parylene-C coating, spincoat a 200 nm thin layer Teflon-AF®3 % w/w for 60 s at 1200 rpm (~32 × *g*) and bake 5 min at 110 °C and 10 min at 200 °C.

3.3 Fabrication of Grounding Plate with Femtoliter-Sized Microwells

1. Clean 2.5 × 2.5 cm^2 cover plates (*see* **Note 6**) by immersing them respectively 5 min in acetone, isopropanol, and deionized water. Repeat this procedure three times and dry the glasses with a nitrogen gun.
2. Put a 2.5 × 2.5 mm^2 piece of copper tape in the center of the glass plates and coat them with a 40 nm thin electrode layer of aluminum via thermal evaporation. After aluminum coating, remove the tape and wash the glasses in acetone, isopropanol, and deionized water, 2 min each, to remove tape residuals (*see* **Note 7**).
3. Treat the aluminum electrode layer for 5 min with O_2 plasma reactive ion etching (100 W, 150 mTorr, 32 sccm O_2) (*see* **Note 8**).
4. Spincoat the glass plates with fluoroalkylsilane Dynasylan F8263 for 15 s at 3000 rpm (~200 × *g*) and let it rest for 1 min. Subsequently, rinse the glass plates with isopropanol and bake for 10 min on a hotplate at 100 °C. The fluoroalkylsilane is used to improve adhesion of Teflon-AF® with the aluminum electrode and glass window.
5. Spincoat the glass plates with Teflon-AF®6 % w/w for 60 s at 500 rpm (~5.6 × *g*) and bake first for 5 min at 110 °C and then 10 min at 250 °C. The resulting Teflon-AF® layer is ≈ 3 μm thick.
6. Apply a thin layer of Parylene-C layer 500 nm via chemical vapor deposition on the top of the Teflon-AF® layer without the use of O_2 plasma and silane (*see* **Note 9**).
7. O_2 plasma activates the Parylene-C mask for 10 s using reactive ion etching (60 W, 100 mTorr, 32 sccm O_2) to improve adhesion with the subsequent aluminum hard mask.
10. Coat a thin aluminum hard-mask layer of ≈ 100 nm via thermal evaporation (*see* **Note 10**).
11. Spincoat adhesion promotor hexamethyldisilazane (HMDS) on the aluminum layer, for 5 s at 500 rpm (~5.6 × *g*) and for 10 s at 3000 rpm (~200 × *g*). Then immediately spincoat AZ1505 positive photoresist for 15 s at 3000 rpm (~200 × *g*) and bake for 2 min on a hotplate at 102 °C.
12. Place the glass plates on the UV-pattern generator, align the microwell array photomask, bring the glass plate and the photomask in close contact, and UV-expose the photoresist at 12 mW/cm^2. Develop the photoresist for 45 s in 351-developer (4× diluted in deionized water) and rinse gently with deionized water. Carefully nitrogen dry the plates (*see* **Note 11**).
13. Check the well diameters using a microscope. They should be 3.5 to 4 μm (*see* **Note 12**).

14. Finally, wet etch the aluminum hard-mask layer by submerging it in aluminum etchant for 3 min at room temperature, rinse with abundant deionized water, and carefully dry with a nitrogen gun (*see* **Note 13**). Again, check the microwells under the microscope to control if the aluminum is completely etched in the bottom of the wells.
15. Dry etch the exposed underlying Parylene-C and Teflon-AF® layers with O_2 plasma reactive ion etching for 10 min (100 W, 150 mTorr, 32 sccm O_2) to obtain 3 μm deep microwells.
16. Carefully peel off the Parylene-C shadow mask together with the aluminum hard mask using a pair of tweezers. The result is a pattern of 2×2 mm^2 hexagonal array of 62,500 femtoliter-sized microwells, with a diameter of 4 μm, a depth of 3 μm, and 8 μm center-to-center spacing, transferred in the Teflon-AF ® layer.

3.4 EWOD Chip Assembly

1. Place the fabricated actuation and grounding plate on a clean wipe tissue.
2. Using a pair of flat tweezers, manually scrape approx. 0.2×0.2 cm^2 wide Teflon-AF® from one of the corners of the grounding plate and apply a silver conductive paint on the exposed electrode. Extend this conductive path from the aluminum layer on the topside toward the grounding plate backside.
3. Using a pair of tweezers, scrap manually the Parylene-C and Teflon-AF® layer deposited on the contacting pads on the periphery of the actuation plate.
4. Clean the grounding and actuation plate with ethanol and water (Milli-Q).
5. Apply two rectangular pieces of double-sided tape (2.0×0.2 cm^2 of a final thickness of 160 μm) on the two opposite edges of the actuation plate.
6. Assemble the actuation plate in the bottom part of the holder (*see* **Note 14**).
7. Using a small volume pipet, aspirate 2.7 μL buffer droplets and place it on an electrode of the actuation plate.
8. Carefully lift the grounding plate from the backside using a vacuum pen, and carefully align it on the top of the actuation plate (Fig. 3). Gently tap the taped sides of the grounding plate using a pair of tweezers (*see* **Note 15**).
9. Pipet 80 μl Vapor-lock™ oil and add them between the assembled DMF plates (*see* **Note 16**).
10. Carefully assemble the top part of the chip holder (*see* **Note 14**).
11. Connect the DMF holder with the PCB board and apply the grounding probe, extended from the PCB board, to the silver paint on the grounding plate.

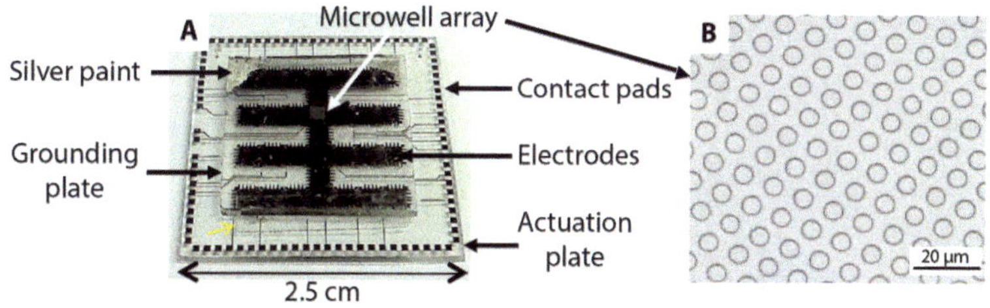

Fig. 3 Schematic overview of the digital microfluidic chip. (**a**) Assembly of grounding and actuation plate. The electrodes are wired to an electrode pad, where the spring loaded connection maintains a contact and provides voltage actuations. The *white arrow* indicates toward a microwell array and the *yellow arrow* indicates toward the applied spacer material. (**b**) The array consists of 62,500 microwells (Ø4.0 µm and 3 µm depth) in an area of 2.0 mm ×2.0 mm

12. Switch on the power supply (frequency generator and amplifier) and the PCB board (*see* **Note 17**).
17. Actuate the droplets using an AC-voltage of 120 to 140 V, an activation time of 300 ms. and a relaxation time of 40 ms. The sequence for actuation of the electrodes can be controlled using homemade MATLAB® and LabVIEW programs.

3.5 Seeding of Superparamagnetic Particles

1. Prepare and assemble the actuation plate in the chip holder as described above (Subheading 3.4)
2. Pipet buffer droplets containing magnetic particles (2.7 µm diameter) on an electrode of the actuation plate, such that after assembly, the droplet aligns with the microwell array of the grounding plate (Fig. 4a) (*see* **Note 18**).
3. Align the microwell containing grounding plate on the top of the actuation plate. Gently tap the sides of the grounding plate using a pair of tweezers.
4. Add 80 µL Vapor-lock™ oil between the assembled DMF plates and fix the top part of the chip holder by applying screws on the four corners. Connect the DMF chip holder with the PCB board and turn on the power supply as described in Subheading 3.4.
5. Position a magnet on the top of the microwell array.
6. Using software controls, actuate the magnetic particle-containing droplet across the microwell array for 10 subsequent seeding cycles (*see* **Note 19**) (Fig. 4b).
7. Transfer the magnetic particle-containing droplet to a separate electrode away from the microwell array.

3.6 Single-Molecule Detection of Biotinylated β-Galactosidase

Whereas the previous paragraph explained how to manipulate magnetic particles on DMF using EWOD, here, a protocol to capture single biotinylated β-Galactosidase molecules (BβG) on

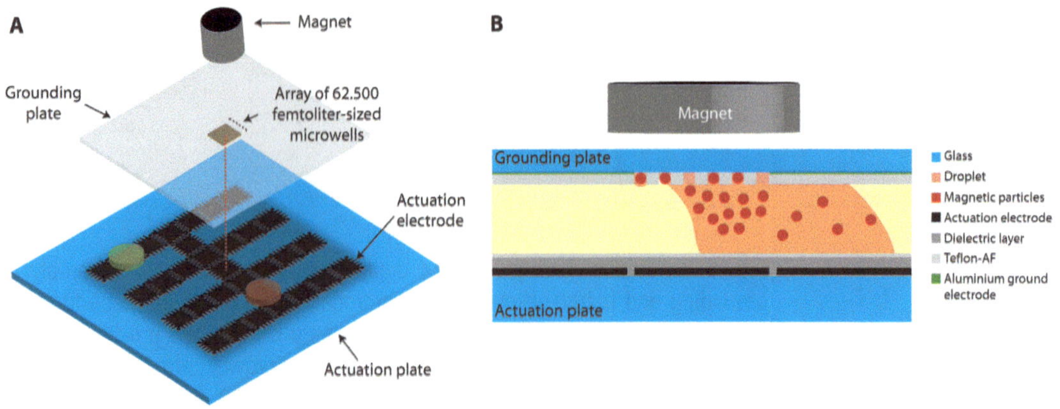

ig.4 High-throughput magnetic particle capturing on the DMF platform. (**a**) Assembly of the DMF platform for capturing single magnetic particles, (**b**) Capturing single magnetic particles inside microwells upon the application of magnetic field in oil environment and performing seeding cycles

streptavidin-coated superparamagnetic particles and to subsequently digitally detect the BβG molecules on the DMF platform is explained.

1. Take 30 μL out of the magnetic particles stock solution (10.5 mg/mL) and wash them three times in washing buffer.
2. Prepare a dilution series of βG molecules (BβG from *Escherichia coli*) ranging from 1.7aM to 170 fM in 400 μL blocking buffer (*see* **Note 20**).
3. Take 100 μL of the particles solution prepared as explained in step 1 and add it to 400 μL blocking buffer containing the different BβG concentrations. Place the samples on a rotational shaker for 4 h at room temperature.
4. After incubation, put the samples in a magnetic rack, remove the supernatant, and wash the particles three times with washing buffer, then three times with PBS containing 0.05 % Pluronic F-68.
5. Resuspend the particles from the different samples in 30 μL of PBS with 0.05 % Pluronic F-68 before manipulating them on the DMF platform.
6. Pipet 2.5 μL of the sample on one actuation electrode of the actuation plate and then sandwich the sample droplet between both plates by putting the grounding plate on top.
7. Shuttle the sample droplets ten times back and forth over the array (*see* previous Subheading 3.6) to seed the magnetic particles in the microwells array.
8. Take 490 μL of PBS buffer and 10 μL of FDG substrate (β-D-Galactopyranoside) to obtain a 100 μM concentration.
9. Pipet a 2.5 μL droplet of FDG in PBS buffer (100 μM) and transport it to the seeded particles using EWOD (*see* Subheadings 3.5 and 3.6).

10. Insert a 2.5 μL Vapor-Lock oil droplet between the DMF plates to avoid evaporation and leakage of the FDG droplet.
11. Move away the FDG from the microwells, thus leaving a femtoliter droplet of FDG inside every single microwell (*see* **Note 21**).
12. Remove the magnet used for increasing the particle seeding efficiency and acquire a first image right after the seeding process of the FDG in the microwells array and use it as a background image (BI). Depending on the microscope objective magnification the number of wells, visible in the field of view, will be different (*see* **Note 22**).
13. Incubate the particles for 30 min with FDG and then capture a second image (I_2).
14. Subtract the background image (BI) taken at 0 min from the one taken at 30 min (I_2) using image analysis software (i.e., ImageJ) to obtain a final subtracted image (I_3).

    Analyze the images with image analysis software (i.e., ImageJ) (*see* **Note 23**) by calculating the percentage of bright spots ("% of active wells") in the subtracted image (I_3) by calculating the ratio between the total number of the particles seeded in the array *NB* (dependent on field of view) (*see* **Note 24**) and the total number of wells which show fluorescence N_{FW} (fluorescent or active wells) as follows:

$$\frac{N_{FW}}{N_B} \times 100\%$$

15. Plot the percentage of active particles vs. the BβG concentration to obtain a graph, as illustrated in Fig. 5a. Figure 5b shows the final subtracted images (I_3) for different concentrations of BβG with next to it their respective seeding efficiency.

4 Notes

1. For the design of the actuation electrodes one can make four identical designs of 3.2 × 3.2 cm², separated by two dicing lines at each side. That way, one can get four plates out of a 4 in. glass wafer during fabrication.
2. When outsourcing photomask fabrication, take into account the fabrication resolution. For the actuation electrode mask the resolution does not need to be really high. However, for the microwell array pattern it is important that the microwell diameter is exactly 3 μm in the final photomask. So for the latter one should make a chrome-on-glass mask using a high resolution LaserPattern generator. In the AutoCAD design, the microwells should be 2.8 μm in diameter to obtain 3 μm in the final photomask after developing and chrome etching.

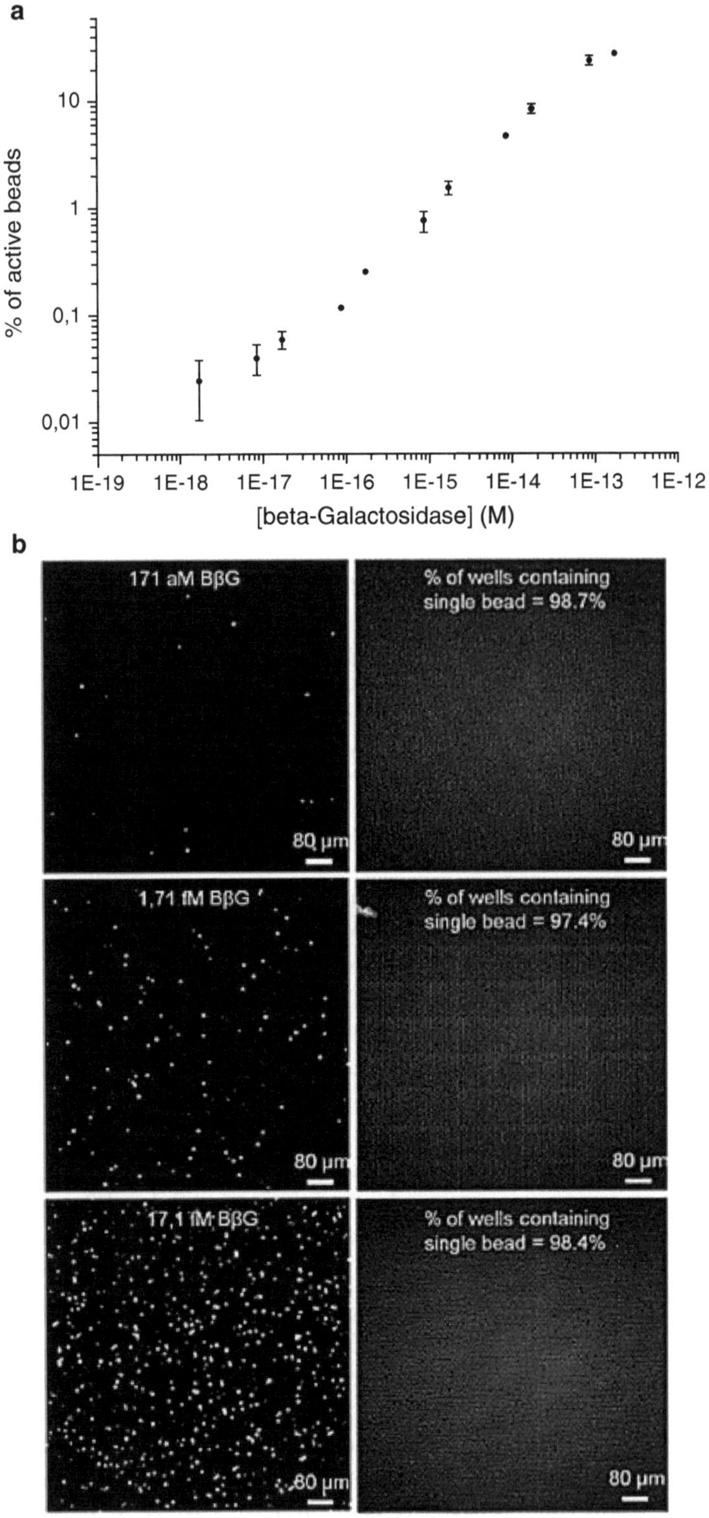

Fig. 5 (**a**) log-log plot of the percentage active particles vs the BβG concentration. Error bars represent standard deviation of three repetitions. (**b**) Subtracted fluorescent images of the digital quantification of BβG on DMF (*left*) and the respective images of particles seeded in the microwell array (*right*)

3. In this case chromium is used as electrode layer and requires a pre-sputtering of 10 min to remove the oxide layer. In case no chromium is available aluminum, silver, platinum, or indium tin oxide can also be used as electrode layer, but these might require different pre-sputtering times.
4. During dicing, align the blade with the dicing marker lines such that the glass wafer is mechanically sawn into different pieces, including the chips. In case a chip does not fit the holder after dicing, then grind the sides of the chip slightly using a diamond sponge scouring pad.
5. Place, together with the plates, a dummy sample in the Parylene-C coating chamber, partially covered with tape. After Parlyene-C deposition remove the tape and measure the layer thickness with a surface profiler.
6. The thickness of the glass cover plates can either be 1.1 mm in case no high resolution microscopy is needed and 0.17 mm thick glass plates can be used in case high magnification objectives need to be used.
7. The aluminum layer should still be transparent such that you can visualize droplet movement afterward under a stereomicroscope.
8. The O_2 plasma reactive ion etching cleans the oxides in the aluminum layer and activates the aluminum dangling bonds.
9. The thin Parylene-C layer serves as a shadow mask that will be removed in the final step so it should not stick to the underlying Teflon-AF® layer.
10. The aluminum layer serves as a hard metal mask necessary to dry etch the microwells with straight walls.
11. The reactivity of 351-developer can cause etching of the underlying aluminum layer in case the photoresist is not coated properly.
12. In case the well diameter is smaller than 3 μm you can extend developing time in steps of 10 seconds. The microwell diameter should definitely not exceed 3.5 or 4 μm.
13. Before using the aluminum etch always turn the bottle carefully upside down a couple of times to mix all the components as they tend to separate during storage.
14. All the experiments were performed on the digital microfluidic platform, using the principle of electrowetting-on-dielectric. The designs and production of a digital microfluidics platform are subject to vary from lab to lab. Various laboratories provide an open-source instruction manual for fabricating custom-designed platform used in their respective labs. Our platform was custom designed and fabricated independently, and comprises of three components, namely actuation software, PCB board,

and a chip holder. The actuation software further is constituted of three elements, a MATLAB®-based user interface, a national instrument-based input software, and an I/O hardware, acquired from by National instrument. The PCB board consists of several discrete circuits, comprising of transistors, diodes, and resistors, where single circuit corresponds to a single electrode port line. The chip holder consists of a top part to provide electrical connections to the electrodes, and a bottom part for holding the digital microfluidic plates.

The actuation software is required as a user interface for accepting and executing the commands defined by the user. The user-defined commands are then processed by an I/O hardware into a transistor-transistor logic (TTL) signal. The TTL signal leads to the activation of a selected circuit on the PCB motherboard that corresponds to the respective electrode actuations, through the spring-loaded electrical connectors, fabricated in our custom-designed chip holder. Activation time refers to the duration for which the electrode is activated with a voltage. The relaxation time is the time between two consecutive electrode activations.

15. During assembly of chips, make sure that the buffer droplets are sandwiched between the hydrophobic surfaces, i.e., Teflon-AF®-coated surfaces.

16. The experiments explained in this chapter were conducted in an oil environment that works great for sealing femtoliter sized droplets in microwell array. However, DMF platform can also be used without the fabrication of microwell array or any special fabricated features on the Teflon-AF® layer of the grounding plate. In such experiments, an oil shell is used to cover the buffer droplet surface, to avoid evaporation. It is done as follows: using a 10 μL pipet, first dip the pipet tip in a low density silicon oil, only to take up the oil due to capillary forces. Then aspirate 2.7 μL buffer droplets in the same pipet tip and place it on an electrode of the actuation plate. The droplet will be contained in an oil shell. While introducing oil between the plates using a pipet, do not push the grounding plate or force the oil abruptly between the plates as it might disrupt the chip assembly. The oil can be neatly introduced by barely placing the pipet tip near the grounding plate edge and releasing the oil. The oil will be taken up between the plates by capillary forces.

17. In our system, an AC voltage was set up using a frequency generator operating at 1 kHz, producing AC voltage in sinusoidal waveforms. The voltage was further amplified using an amplifier to reach the required final voltage.

18. The droplet has to be added on an electrode in such a way that after the DMF plate assembly, the microwell array overlaps with the droplet. In case the droplet is not aligned properly

upon assembly, disassemble the chips, relocate the magnetic particle containing droplet, and align the microwell array containing grounding plate.

19. The seeding cycle refers to every one-way transportation of the droplet across the microwell array. The seeding of magnetic particles can also be done manually by simply placing a magnet below the array and transporting the droplet, with suspended magnetic particles, back and forward using a pipet tip. However, microwell array containing digital microfluidic platform is required for conducting the complete digital assay in an automated manner.
20. In order to get one single molecule per magnetic particle, the ratio between the concentration of the particles and the concentration of the (BβG) molecules is established to be between zero and one according to the Poisson distribution concept. Hence, by playing with the concentrations, the certainty that every particle carries maximum one BβG molecule on its surface can be achieved and bioassays can be performed in the digital regime.
21. In this configuration, every single microwell of the array contains a droplet of FDG in which one single-seeded particle is immersed. In the presence of the BβG enzyme molecule on a particle, a fluorogenic reaction starts in the microwell containing the particle. The FDG substrate is directly converted into a fluorescent product only in those wells where the βG enzyme is present because the latest first hydrolyzes nonfluorescent FDG to fluorescein monogalactoside (FMG) and then to highly fluorescent fluorescein. Moreover, all the microwells are encapsulated in the oil matrix, thus avoiding any possible leak and evaporation of the FDG substrate.
22. All the images are acquired using a microscope equipped with a CCD camera and a 20× objective, which results in a field of view of around 10,000 microwells. This image area is sufficiently large to do statistical analysis of the generated signal. The number of imaged microwells thus depends on the field of view and can be increased by either using a lower magnification objective or by scanning the whole microwell array area and stitching the images. In the second case, 100 % of the microwells can be imaged. In order to acquire images showing optimal balance between the brightness of the spots and the dark background, the exposure time of the camera is established at 1 s. This exposure time was set by referring to a control sample. Too low or high exposure times may result respectively in dark or bright images which cannot be analyzed with the image analysis software (i.e., ImageJ).
23. All the bright spots in the images can be easily counted by using the "Analyze particles function" in Image J software (US 0National Institutes of Health, http://rsb.info.nih.gov/iJ/).

24. Calculate N_B: since Lodestar particles show autofluorescence (excitation 530–550 nm and emission 575–625 nm), N_B can be calculated by exposing the array with the seeded particles to the appropriate intensity light and visualize it under the fluorescent microscope.

Acknowledgments

This research was financially supported by the KU Leuven Research Council (IDO-project 10/012, OT project 13/058 and Atheromix IOF-knowledge platform), the Agency for Innovation by Science and Technology in Flanders (IWT project 121615), and the Fund for Scientific Research Flanders—FWO (G.0997.11 and G.0861.14).

References

1. Rondelez Y, Tresset G, Nakashima T (2005) Highly coupled ATP synthesis by F 1-ATPase single molecules. Nature 433:773–777
2. Rissin DM, Walt DR (2006) Digital concentration readout of single enzyme molecules using femtoliter arrays Poisson statistics. Nano Lett 6:520–523
3. Rissin DM, Walt DR (2006) Digital readout of target binding with attomole detection limits via enzyme amplification in femtoliter arrays. J Am Chem Soc 128:6286–6287
4. Rissin D, Kan C, Campbell T (2010) Single-molecule enzyme-linked immunosorbent assay detects serum proteins at subfemtomolar concentrations. Nat Biotechnol. 28:595–589
5. Kan CW, Rivnak AJ, Campbell TG, Piech T, Rissin DM, Mösl M et al (2012) Isolationdetection of single molecules on paramagnetic beads using sequential fluid flows in microfabricated polymer array assemblies. Lab Chip 12:977–985
6. Zhang H, Nie S, Etson CM, Wang RM, Walt DR (2012) Oil-sealed femtoliter fiber-optic arrays for single molecule analysis. Lab Chip 12:2229–2239
7. Kim SH, Iwai S, Araki S, Sakakihara S, Iino R, Noji H (2012) Large-scale femtoliter droplet array for digital counting of single biomolecules. Lab Chip 12:4986–4991
8. Chang L, Rissin DM, Fournier DR, Piech T, Patel PP, Wilson DH et al (2012) Single molecule enzyme-linked immunosorbent assays: theoretical considerations. J Immunol Methods 378:102–115
9. Witters D, Knez K, Ceyssens F, Puers R, Lammertyn J (2013) Digital microfluidics-enabled single-molecule detection by printing and sealing single magnetic beads in femtoliter droplets. Lab Chip 13:2047–2054
10. Pollack MG, AD S, RB F (2002) Electrowetting-based actuation of droplets for integrated microfluidics. Lab Chip 2:96–101
11. Fair RB (2007) Digital microfluidics: is a true lab-on-a-chip possible? Microfluid Nanofluid 3:245–281
12. Abdelgawad M, Wheeler AR (2009) The Digital Revolution: A New Paradigm for Microfluidics. Adv Mater 21:920–925
13. Witters D, Vergauwe N, Vermeir S, Ceyssens F, Liekens S, Puers R et al (2011) Biofunctionalization of electrowetting-on-dielectric digital microfluidic chips for miniaturized cell-based applications. Lab Chip 16: 2790–2794
14. Vergauwe N, Witters D, Atalay YT, Verbruggen B, Vermeir S, Ceyssens F et al (2011) Controlling droplet size variability of a digital lab-on-a-chip for improved bio-assay performance. Microfluid Nanofluid 11:25–34
15. Vergauwe N, Witters D, Ceyssens F, Vermeir S, Verbruggen B, Puers R et al (2011) A versatile electrowetting-based digital microfluidic platform for quantitative homogeneous and heterogeneous bio-assays. J Micromechanics Microengineering 21:054026
16. Kühnemund M, Witters D, Nilsson M, Lammertyn J (2014) Circle-to-circle amplification on a digital microfluidic chip for amplified single molecule detection. Lab Chip 16: 2983–2992
17. KP T, Toffalini F, Witters D, Vermeir S, Rolland F, MLATM H et al (2014) Digital

microfluidic chip technology for water permeability measurements on single isolated plant protoplasts. Sensors Actuators B Chem 199:479–487

18. Witters D, Vergauwe N, Ameloot R, Vermeir S, De Vos D, Puers R et al (2012) Digital microfluidic high-throughput printing of single metal-organic framework crystals. Adv Mater 24:1316–1320
19. Tewari KP, Vriens K, Cornaglia M, Gijs M, Kokalj T, Thevissen K et al (2015) Digital microfluidics for time-resolved cytotoxicity studies on single non-adherent yeast cells. Lab Chip 8:1852–1860
20. Kokalj T, Pérez-Ruiz E, Lammertyn J (2015) Building bio-assays with magnetic particles on a digital microfluidic platform. N Biotechnol 32(5):485–503

Part III

Microchips for Nucleic Acids Analysis

Chapter 8

Microfluidic DNA Stretching Device for Single-Molecule Diagnostics

Daisuke Onoshima and Yoshinobu Baba

Abstract

The method described here enables the automatic stretching and patterning of single DNA molecules onto a solid surface. It does not require chemical modification of the DNA or surface modification of the substrate. To detect a signal variation caused by sequence-specific dye binding or partial melting, it is crucial that the DNA molecules are arrayed in a parallel direction inside the narrow microscopic field. The method uses zigzag-shaped microgrooves in a densely-arranged molecular patterning apparatus in a microfluidic channel. By syringing through the microchannel, over 1500 DNA molecules can be arrayed simultaneously in the microgrooves. It will therefore serve as a template preparation for DNA molecular diagnosis by high-resolution imaging.

Key words Microfluidic device, Single-Molecule detection, DNA stretching, Optical mapping, Molecular diagnosis

1 Introduction

Positioning DNA on a solid surface is a crucial technique for genetic and epigenetic mapping by optical detection [1–4]. The DNA molecule in solution generally adopts an entropically favorable random-coil conformation. By applying tension to the strand, the DNA molecule is stretched into an entropically unfavorable elongated state. In the case of aligning terminal-unmodified DNA in a single direction on a surface, molecular combing by dewetting the DNA solution [5–13] is widely used. This method uses the surface tension force at the air–water interface of the DNA solution translating on the glass substrate. When the speeds of molecular translation and adsorption are balanced, the DNA is stretched along the direction of the liquid interface.

The capability of a DNA molecular stretching can be maximized when the bottom surface of the substrate is microfabricated. By specifying the location of DNA attachment on the fabricated patterns, the substrate can serve as a template that can assist in

Valérie Taly et al. (eds.), *Microchip Diagnostics: Methods and Protocols*, Methods in Molecular Biology, vol. 1547,
DOI 10.1007/978-1-4939-6734-6_8, © Springer Science+Business Media LLC 2017

high-resolution imaging studies such as atomic force microscopy [14] and transmission electron microscopy [15]. The mechanism for DNA trapping is driven by the receding meniscus assembly and the capillary force from the dewetting process [16, 17]. However, the procedure recently used for molecular stretching still does not have controls to handle liquid solutions. In terms of this kind of liquid handling, a microfluidic device [18, 19] is expected to have advantages over the conventional method. In this chapter, we describe a microfluidic device for stretching terminal-unmodified single DNA molecules by liquid sucking [20]. The liquid contact line inside the microfluidic channel is manipulated by pressure control. A microfabricated pattern for trapping single DNA molecules is embedded onto the surface of poly(dimethylsiloxane) (PDMS) microfluidic channel. It comprises two types of PDMS substrates: a PDMS microchannel for DNA solution transport and a zigzag-shaped PDMS microgroove for DNA trapping. The development of this device thus includes the microfabrication of the zigzag pattern on the PDMS surface for a high-density DNA array.

2 Materials

All solutions were kept in a refrigerator or a freezer. Frozen samples were thawed in a refrigerator just before use.

2.1 Photomask

1. Glass mask blank with a Cr layer and a photoresist layer (CBL4009Bu-AZP; Clean Surface Technology, Japan).
2. Developing solution (NMD-3; Tokyo Ohka Kogyo, Japan). Store at 4 °C.
3. Cr etchant solution: 17 g of ammonium second cerium nitrate, 5 mL of perchloric acid, and 100 mL of purified water. Store at 4 °C.
4. Stripping solution (Hakuri 104, Tokyo Ohka Kogyo, Japan). Store at 4 °C.
5. Maskless exposure machine (NanoSystem Solutions, Japan).
6. Hot plate (Asone, Japan).
7. CAD software (AutoCAD, Autodesk, USA).

2.2 Photoresist and Silicone

1. SU-8 photoresist (SU-8 3050 and SU-8 3005, Nippon Kayaku, Japan). Store at 4 °C.
2. SU-8 developer (Nippon Kayaku, Japan). Store at 4 °C.
3. Isopropyl alcohol. Store at 4 °C.
4. PDMS prepolymer solution: 10:1 mass ratio of PDMS oligomers (SILPOT 184 Dow Corning, Toray, Japan) and a reticular agent (Catalyst SILPOT 184 Dow Corning, Toray, Japan). Store at 4 °C (*see* **Note 1**).

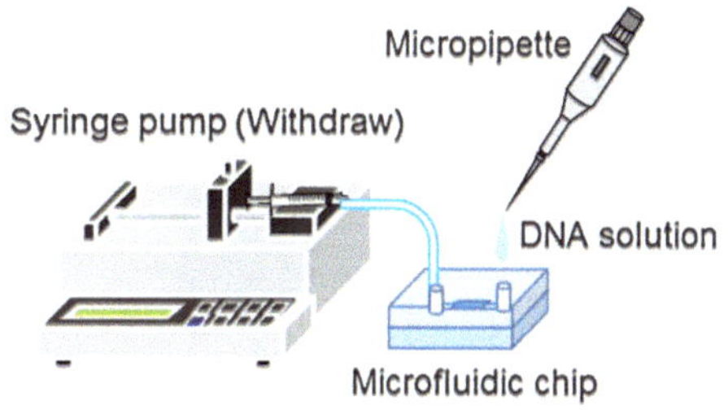

Fig. 1 Schematic of microfluidic DNA molecular stretching device

5. Silanization agent: Trichloro (1H, 1H, 2H, 2H-perfluorooctyl) silane (Sigma-Aldrich, Canada). Store at 4 °C.
6. 4-inch silicon substrate (Ferrotec, Japan).
7. Spin coater (IF-D7, Mikasa).
8. Mask aligner (M-1S, Mikasa, Japan).
9. Surgical scalpel blade.
10. Uni-core stainless steel cutting tip (Harris, USA).

2.3 DNA, Pumping System, and Microscope

1. Lambda DNA (48.5 kbp): 0.2–0.8 μg/μL in 10 mM Tris–HCl (pH 8.0) and 1 mM EDTA. Store at −20 °C (*see* **Note 2**).
2. Fluorescent dye solution: 1 mM YOYO-1 iodide in DMSO. Store at −20 °C (*see* **Note 3**).
3. TE Buffer: 10 mM Tris–HCl (pH 8.0) and 1 mM EDTA. Store at 4 °C.
4. Syringe pump (KDS-200, Kd Scientific, USA) with a Hamilton syringe (500 μL) and a silicon tube (Fig. 1).
5. Total internal reflection fluorescence microscope (IX2-RFAEVA-2, Olympus, Japan) with an EM-CCD camera (Hamamatsu Photonics, Japan).

3 Methods

All procedures were carried out at room temperature unless otherwise specified.

3.1 Mask Patterning

1. Draw a microfluidic channel pattern (650 μm width, 0.8 cm length, and 50 μm height) for DNA solution transport and 16 zigzag-shaped microgroove patterns (Fig. 2) for DNA trapping on the CAD software (*see* **Note 4**).
2. Input the CAD data into the maskless exposure machine.
3. Exposure the glass mask blank by using the maskless exposure machine. Set the exposure energy to 100 mJ/cm^2.

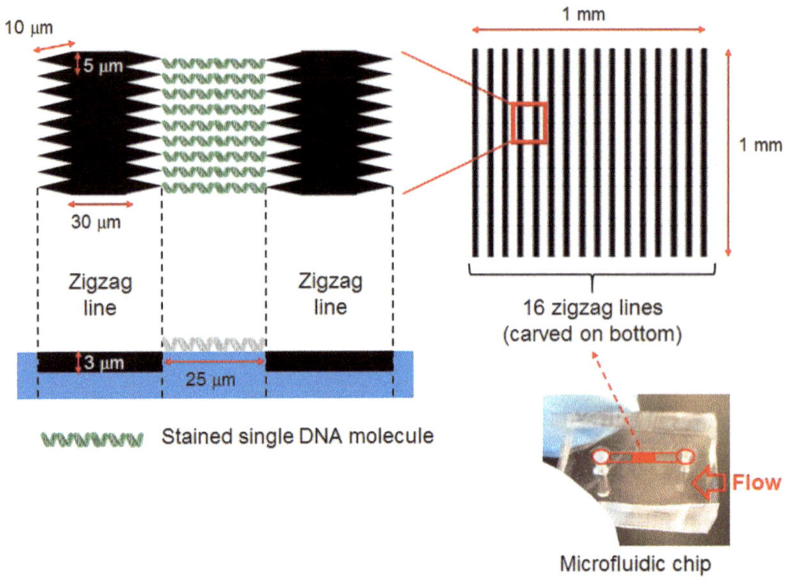

Fig. 2 Detailed explanation of the microfluidic chip and the pattern for DNA trapping

4. Develop the glass mask by immersing it in a developing solution for 1 min.
5. Wash the glass mask by rinsing with pure water.
6. Etch the exposed Cr layer on the glass mask in Cr etchant solution for 1 min.
7. Wash the glass mask with the stripping solution.
8. Bake the glass mask for 20 min at 80 °C on the hot plate.

3.2 Soft Lithography

1. Drop the SU-8 photoresist onto the silicon substrate and coat the substrate surface with the photoresist by using the spin coater (*see* **Note 5**).
2. Bake the substrate for 20 min at 95 °C on the hot plate.
3. Expose the substrate to UV light through the patterned glass mask by using the mask aligner. Set the exposure energy to 250 mJ/cm^2.
4. Bake the substrate for 5 min at 95 °C on the hot plate.
5. Immerse the substrate in the SU-8 developer for 8 min (*see* **Note 6**).
6. Wash the substrate by rinsing with the SU-8 developer and isopropyl alcohol.
7. Place the substrate into a desiccator for 60 min with several drops of the silanization agent under vacuum.
8. Pour the PDMS prepolymer solution onto the substrate in a plastic dish.

9. Place it into a desiccator and expel the remaining air for 2 h under vacuum.
10. Bake it for 2 h at 80 °C in an oven.
11. Peel off the PDMS substrate and cut by using surgical scalpel blade.
12. Punch an inlet hole and an outlet hole as reservoirs of the PDMS microchannel by using the uni-core stainless steel cutting tip. The sizes of the inlet and outlet holes are 1.5 and 1 mm in diameter, respectively.
13. Bond the flow path of the PDMS microchannel with the zigzag pattern of the PDMS microgroove (*see* **Note 7**).

3.3 DNA Stretching and Patterning

1. Stain the lambda DNA with the YOYO-1 intercalating dye at a 1:5 dye to base pair ratio overnight in the dark at 4 °C. Dilute the stained DNA solution to 10 ng/μL with TE buffer.
2. Connect the outlet hole of the PDMS microchannel to the syringe pump via the Hamilton syringe through the silicon tube.
3. Set the flow rate to 1 μL/min.
4. Drop 10 μL of TE buffer into the inlet hole and suck through the microchannel for 30 min.
5. Drop 10 μL of the DNA solution into the inlet hole and suck through the microchannel for 30 min (*see* **Note 8**).
6. Peel off the zigzag-shaped PDMS microgroove and bring into contact with a cover glass for microscopic observation (*see* **Note 9**).
7. Observe and count the trapped single DNA molecules in the microgrooves by using the total internal reflection fluorescence microscope (*see* **Note 10**) (Fig. 3).

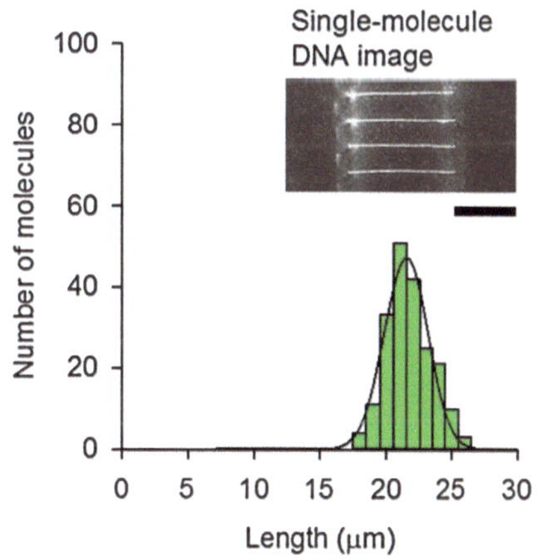

Fig. 3 Results of trapping and sizing by single DNA molecules. The histogram consists of 200 DNA molecules obtained from the fluorescence image analysis, and the also shown line is the Gaussian fit. The scale bar is 10 μm

4 Notes

1. The PDMS prepolymer solution can be mixed uniformly by using a centrifugal (2000 rpm for 2 min, then 2200 rpm for 6 min). Degassing also takes place during this centrifugal treatment.
2. In order to avoid the deleterious effect of repeated freeze/thaw cycles, it is recommended that samples should be aliquoted in small volumes (20 μL) and then stored frozen. For example, a DNA cleavage might be due to the repeated freeze/thaw cycles.
3. After preparation, shield the solution from light with aluminum foil.
4. The zigzag pattern is constructed using line and triangle patterns in a 1 × 1 mm area. The line width is 30 μm and the length is 1 mm. The triangle patterns are combined with the edges of lines. Patterns using an isosceles triangle (base length of 5 μm and leg length of 10 μm) are used. The distance between the zigzag vertices (25 μm) can be changed for optimal DNA size.
5. Use the following two spin coating protocols for the PDMS microchannel and the PDMS microgroove; (1) The thickness of 50 μm (SU8-3050) for the PDMS microchannel: slope for 5.0 s, 500 rpm for 10 s, slope for 7.0 s, and 2800 rpm for 30 s. (2) The thickness of 3 μm (SU-8 3005) for the PDMS microgroove: slope for 5.0 s, 500 rpm for 10 s, slope for 7.0 s, and 4000 rpm for 30 s.
6. During the incubation, the substrate should be shaken.
7. The surfaces of the two PDMS structures can be glued together by self-adsorption.
8. Keep the sucking for 30 min. The Arrayed polymer chains of DNA are fixed by drying in this procedure.
9. The arrayed DNA molecules can be transferred to a silanized glass by the following procedures: (1) Immerse a cover glass in 0.1 M NaOH for 1 min, and subsequently in 99 % ethanol for 15 min; (2) Clean the cover glass ultrasonically in ethanol for 10 min, and immerse in (3-Aminopropyl)triethoxysilane (APTES) solution for 1 h. The APTES solution contains APTES, 99 % ethanol, and glacial acetic acid in a ratio of 30:970:1; (3) Wash the cover glass in 99 % ethanol for 20 min, and heat for 2 h at 120 °C on a hot plate; (4) Bring the cover glass into contact with the trapped DNA molecules on the PDMS microgroove for 1 min within 24 h of the silanization.
10. The single zigzag line has 200 corners of triangles. The central 100 corners of triangles in the 15 zigzag lines are effective for image analysis.

Acknowledgment

This work was supported by JSPS KAKENHI Grant Number 26750146. This research is partly supported by the Japan Society for the Center of Innovation Program from Japan Science and Technology Agency, JST.

References

1. Neely RK, Deen J, Hofkens J (2011) Optical mapping of DNA: single-molecule-based methods for mapping genomes. Biopolymers 95:298–311
2. Kim S et al (2012) Enzymatically incorporated genomic tags for optical mapping of DNA-binding proteins. Angew Chem Int Ed 51:3578–3581
3. Michaeli Y et al (2013) Optical detection of epigenetic marks: sensitive quantification and direct imaging of individual hydroxymethylcytosine bases. Chem Commun 49:8599–8601
4. Cerf A, Cipriany BR, Benitez JJ, Craighead HG (2011) Single DNA molecule patterning for high-throughput epigenetic mapping. Anal Chem 83:8073–8077
5. Lebofsky R, Bensimon A (2003) Single DNA molecule analysis: applications of molecular combing. Brief Funct Genomic Proteomic 1:385–396
6. Michalet X et al (1997) Dynamic molecular combing: stretching the whole human genome for high-resolution studies. Science 277: 1518–1523
7. Gueroui Z, Place C, Freyssingeas E, Berge B (2002) Observation by fluorescence microscopy of transcription on single combed DNA. Proc Natl Acad Sci USA 99: 6005–6010
8. Petit CAP, Carbeck JD (2003) Combing of molecules in microchannels (COMMIC): a method for micropatterning and orienting stretched molecules of DNA on a surface. Nano Lett 3:1141–1146
9. Bensimon A et al (1994) Alignment and sensitive detection of DNA by a moving interface. Science 265:2096–2098
10. Bensimon D, Simon AJ, Croquette V, Bensimon A (1995) Stretching DNA with a receding meniscus: experiments and models. Phys Rev Lett 74:4754–4757
11. Allemand JF et al (1997) pH-dependent specific binding and combing of DNA. Biophys J 73:2064–2070
12. Hu J et al (1996) Imaging of single extended DNA molecules on flat (aminopropyl) triethoxysilane-mica by atomic force microscopy. Langmuir 12:1697–1700
13. Woolley AT, Kelly RT (2001) Deposition and characterization of extended single-stranded DNA molecules on surfaces. Nano Lett 1:345–348
14. Aline C, Thomas A, Barton RA, Craighead HG (2011) Transfer-printing of single DNA molecule arrays on graphene for high-resolution electron imaging and analysis. Nano Lett 11:4232–4238
15. Gentile F et al (2012) Direct imaging of DNA fibers: the visage of double helix. Nano Lett 12:6453–6458
16. Lin CJ et al (2010) Patterning nanowire and micro-nanoparticle array on micropillar-structured surface: Experiment and modeling. Biomicrofluidics 4:034103
17. Charlot B et al (2014) Elongated unique DNA strand deposition on microstructured substrate by receding meniscus assembly and capillary force. Biomicrofluidics 8:014103
18. Sia SK, Whitesides GM (2003) Microfluidic devices fabricated in poly(dimethylsiloxane) for biological studies. Electrophoresis 24:3563–3576
19. Kaji N, Okamoto Y, Tokeshi M, Baba Y (2010) Nanopillar, nanoball, and nanofibers for highly efficient analysis of biomolecules. Chem Soc Rev 39:948–956
20. Yasaki H et al (2015) Microfluidic transfer of liquid interface for parallel stretching and stamping of terminal-unmodified single DNA molecules in zigzag-shaped microgrooves. Lab Chip 15:135–140

Chapter 9

Droplet Microfluidic and Magnetic Particles Platform for Cancer Typing

Davide Ferraro, Jérôme Champ, Bruno Teste, M. Serra, Laurent Malaquin, Stéphanie Descroix, Patricia de Cremoux, and Jean-Louis Viovy

Abstract

Analyses of nucleic acids are routinely performed in hospital laboratories to detect gene alterations for cancer diagnosis and treatment decision. Among the different possible investigations, mRNA analysis provides information on abnormal levels of genes expression. Standard laboratory methods are still not adapted to the isolation and quantitation of low mRNA amounts and new techniques needs to be developed in particular for rare subsets analysis. By reducing the volume involved, time process, and the contamination risks, droplet microfluidics provide numerous advantages to perform analysis down to the single cell level.

We report on a droplet microfluidic platform based on the manipulation of magnetic particles that allows the clinical analysis of tumor tissues. In particular, it allows the extraction of mRNA from the total-RNA sample, Reverse Transcription, and cDNA amplification, all in droplets.

Key words Droplet microfluidics, Magnetic tweezers, Reverse Transcription (RT), PCR

1 Introduction

Microfluidics and especially droplet microfluidics are revolutionizing the typical nucleic acids analysis [1]. As a matter of fact, these new technologies allow reducing the contamination risks, the reagents volume involved, and so the relative costs, as well as increasing the sensitivity compared with conventional strategies. Additionally, the reduction in the amount of reagents of up to one million-fold makes experiments that were previously prohibitively expensive feasible, and allows assays to be performed on systems that are not available in large quantities. Therefore, miniaturization makes this approach ideally suited for point-of-care applications in diagnostics, clinical management, and personalized medicine. Pioneer works have already shown the potential of microfluidics for transcriptome analysis [2, 3]. However, they were limited to "housekeeping" genes presenting high levels of

Valérie Taly et al. (eds.), *Microchip Diagnostics: Methods and Protocols*, Methods in Molecular Biology, vol. 1547,
DOI 10.1007/978-1-4939-6734-6_9,

expression and no pathological relevance. Abate et al. [4] showed the possibility of performing Reverse Transcription (RT) and polymerase chain reaction (PCR) on total-RNA sample, starting from cancer cell lines in droplet microfluidic devices. The final distribution of the droplet in tubes, however, increases drastically the risks of droplet breaking and RNAse contamination. Higher programmability was allowed by electro-wetting on dielectric technology (EWOD), and a device performing mRNA extraction, purification, and transcription has been recently described [5]. EWOD, however, requires strong microfabrication expertise and expensive facilities, and also requires additional upstream interface for sample and reagents loading.

Here, we present a microfluidic platform that allows extraction of mRNA from a total RNA sample using magnetic beads as solid support, washing, resuspension in new solution, and Reverse Transcription and amplification, all in droplets.

Nowadays, there are different strategies for droplet merging, mixing, and splitting which allow several types of operations and reactions that are typically performed at the microscale [6]. Notably, however, one operation missing in this list of possibility is sample purification. This represents a limitation because most of the strategies for nucleic acids and genomic analysis require extraction and purification steps commonly performed with magnetic particles. We recently overcome this bottleneck, thanks to the development of a magnetic tweezers technology [7, 8]. Colloidal suspension of magnetic particles functionalized with capture target is injected in the droplet; applying an external magnetic field, the particles can be confined into a cluster that can be extracted from the droplet in a very small volume and merged with an incoming droplet. This technology is particularly adapted to diagnosis and single cells applications, in which a moderate number of samples or cells (typically from ten to a few hundreds) have to be screened. Combining the magnetic tweezers technology with a pipetting robot for droplet generation, it has been possible to create an automated platform completely compatible with the conventional equipment present in biological laboratory, like a microtiter plate [9]. More in detail, this platform allows the production of controlled "trains" of confined droplets, in which each droplet contains a different, deterministically programmed, sample, reagent or buffer from independent libraries. Additionally, the droplet generation from a conventional microtiter plate makes the platform compatible with the conventional equipment of the biological laboratory and suitable for the typical protocol, like performing triplicate experiments on the same sample, negative control or different combination between samples and biomarkers to study. As a proof of concept, the microfluidic platform was validated by the screening of two genes of interest: TPB (TATA-binding protein) and HER2 (Human Epidermal Growth Factor

Receptor-2), starting from MCF7 and SKBR3 cell lines. The first gene is a housekeeping gene, typically used as a reference in clinical laboratory, while the second is an important biomarker for breast cancer [10], which represents an emblematic example of the concept of personalized medicine. This gene is involved in the regulation of cell growth and proliferation and it has been reported in different cancer types (breast, ovarian, stomach). Additionally, its overexpression (HER2+ status) is strongly associated with increased disease recurrence and with a poor prognosis. Nowadays, several targeted therapies, in particular based on monoclonal antibodies targeting the HER-2 receptor (Trastuzumab), have shown strongly improved survival for patients with HER2 overexpression. This treatment, however, is ineffective in metastatic breast cancer patients with a HER2-negative status and could be associated with an increased risk of cardiotoxicity [11]. Therefore, a reliable evaluation of HER2 expression level is determinant in the daily practice of clinicians.

2 Materials

All the solutions are prepared at room temperature in a pre-PCR room just before placing in a 384-well microtiter plate for the droplet generation.

2.1 RT-qPCR and Droplet Solutions

1. 100 μL of Dynabeads Oligo(dT)25 Kit (Life Technologies) is concentrated four times in 25 μL of the Binding buffer solutions present in the Kit. 0.5 μL of BSA at 20 % is added.
2. Total RNA, stored at −80 °C, is quantified and qualified using Nanodrop & BioAnalyzer (Agilent) instruments and then, it is diluted in 25 μL of binding buffer (Dynabeads Kit) to obtain the desired concentration.0.5 μL of BSA at 20 % is added.
3. 25 μL of washing buffer solution (Dynabeads Kit) is mixed with 0.5 μL of BSA at 20 %.
4. CellsDirect™ One-Step qRT-PCR kits (Life Technologies) was used to prepare the Reverse transcription mix as follows: 12.5 μL of 2× Reaction Mix, 1 μL of anchored Oligo(dT)23VN (S1327S, NEB), 1 μL of each Forward and Reverse primers (custom primers to study HER2 or TBP or Actin b from IDT), 1 μL of SuperScript® III RT/Platinum® Taq Mix with RNaseOUT™ Ribonuclease Inhibitor, and enough quantity of DEPC-treated water to reach 25 μL.
5. After the microfluidic process (mRNA captured, washing, Reverse transcription step), droplets containing cDNA were recovered into 200 μL tubes with 5 μL of DEPC-treated water. qPCR reactions were done into Smartcycler® instrument (Cepheid), with a mix composed of: 12.5 μL of 2×

Reaction Mix, 1 μL of each Forward and Reverse primers (custom primers to study HER2 or TBP or Actin b from IDT), 0.5 μL of TaqMan™ Probe (custom probes to study HER2 or TBP or Actin b from IDT), 1 μL of SuperScript® III RT/Platinum® Taq Mix with RNaseOUT™ Ribonuclease Inhibitor, 5 μL of cDNA output and qsp to 25 μL with DEPC-treated water. The thermocycling program was: an enzyme activation step at 95 °C for 120 s, followed by 50 cycles of 95 °C/15 s and 60 °C/30 s.

2.2 Droplet Microfluidic Platform

1. The droplet generation is achieved by an arm-robot (Rotaxys, Cetoni) connected with two syringe pumps (Nemesys, Cetoni) which moves a PTFE tubing (inside/outside diameter: 0.3 mm/0.6 mm) between the different wells of the microtiter plate. To space the water phase droplets, FC-40 + 2 % surfactant (1H,1H,2H,2H-perfluorodecan-1-ol, Fluorochem) is used as a continuous phase. The robot and the syringe pumps are controlled by a script generated by QmixElement software (cetoni GmbH). As shown in Fig. 1, the syringe pumps are equipped with 2 SGE syringes (250 μL and 2.5 mL), which are connected with two PDMS chips and two pinch valves (NResearch Inc.). The PDMS chips are prepared by conventional soft-lithography and the inside channels have a size of 200 × 200 μm. In detail, the syringe 1 is connected with the inlet 1 of the PDMS chip 1 (Fig. 1b), while the syringe 2 is connected with the other PDMS chip 2 (Fig. 1c) at the inlet 4; the inlets 2 and 3 are connected respectively at the robot and at the other PDMS device (inlet 5), through the two valves (V1 and V2). Finally, the inlet 6 of the second PDMS chip is connected with the capillary part which brings the droplets at the magnetic tweezers. This system guarantees droplet generation while other droplets are already flowing in the rest of the platform. The PDMS devices are prepared by standard double replica molding technique and silicon tubing (0.31/0.64 mm ID/OD, Helix Medical) is used in the pinch valves. The connection between the different tubing is assured by a third silicon capillary having 0.50 mm as internal and 1.3 as external diameter, respectively (Deutsch & Neumann). To assure the good wettability conditions, PDMS devices and silicon capillaries were treated with a solution of (Tridecafluoro-1,1,2,2-tetrahydrooctyl)trichlorosilane in FC-40 (5 %) for 1 h.
2. The magnetic tweezers (Fig. 1d) are composed by two magnetic coils (copper wire of 71 cm and 1 mm diameter, Radiospares) within paramagnetic tips (AFK502 alloy, Aperam); therefore turning ON and OFF the current, the magnetic field is controlled. For the droplet observation and the magnetic tweezers control two Macro Objectives (MLH-10X) mounted on two CMOS Cameras (acA1300-60gm, by Basler)

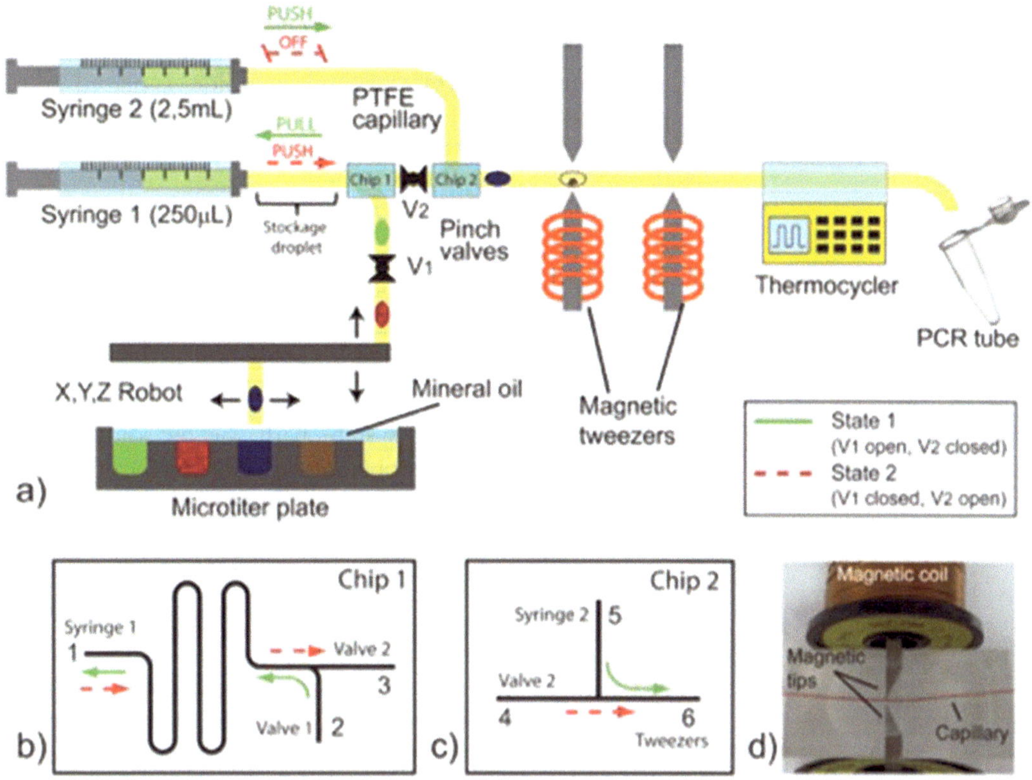

Fig. 1 (**a**) Scheme of the droplet microfluidic platform composed by three main parts: the droplet module generator, the magnetic tweezers, and the thermocycler. The droplet generator module combines a microtiter plate placed under an arm-robot, two syringes, two pinch valves, and two PDMS chips (**b**–**c**). First, the solutions for droplets generation are prepared in a microtiter plate. Syringe pumps and valves are software controlled, making the system completely automatize. It continuously switches between states 1 and 2, represented by *green continuous line* and *red dashed line*, respectively. During state 1, V1 is open and V2 is closed, one train of droplets is generated, passing through V1 and finally stored in the serpentine of the PDMS chip 1 while, at the same time, the syringe 2 pushes the oil in the rest of the capillary. After that, system switches to State 2 (V1 closed and V2 open): syringe 2 is OFF and syringe 1 pushes the droplet beyond V2 and the PDMS chip 2. Finally, State 1 is reactivated for the next train. In this way, the system can work continuously: when one syringe is generating the droplet train, the other is driving the rest of the droplets through the next modules of the platform. A mineral oil film of 2–3 mm is spread on the microtiter plate avoiding contaminations or solution evaporations; this is kept by a plastic wall bonded on the plate. After their generation, droplets are driven between two magnetic tweezers (**d**) composed by magnetic coils and paramagnetic tips, which can be magnetized activating the coils. The capillary colored in *red* in the Figure is placed in close contact with one of the tips. After the magnetic beads manipulation, droplets are driven to the adapted thermocycler for the RT

are used. The cameras are connected at the computer by the Frame Grabber NI PCIe-8236 (National Instruments) and the communication between the home-made LabVIEW software and the magnetic coils is achieved by the NI USB-6525 (National Instrument). The software is able to detect the beads and the droplet meniscus, allowing the current actuation in the magnetic coil.

3. Techne Prime thermocycler equipped with Techne in situ Hybridization Adapter (flat plate for glass slides) is used for the RT. An aluminum part with a capillary holder is placed on the flat plate and the PTFE tubing is fixed in it.

3 Methods

All the steps of the procedure are performed at room temperature and, a part where specified, in a conventional physical laboratory, without particular restriction concerning the biological environment.

1. Fill all the tubing of the system with the oil solution (FC-40 + 2 % surfactant). This operation can be performed in two ways: pushing the oil stored in the syringes, or inspiring it from the open side of the tubing, after the thermocycler. Both strategies are equivalent; do not trap air bubbles during this step.
2. Switch the thermocycler ON and fix the temperature at 50 °C.
3. Set the script selecting the wells where the solutions for the droplet generation will be placed (water phase and oil). Since the oil is sucked more than the water phases, it is better to envisage one well full of oil every five trains of droplets. Set the flow-rate for driving the trains of droplets in the tubing at 0.1 μL/s
4. Prepare the water phase solutions described in the paragraph 2.1 in 0.2 mL tube in a pre-PCR room.
5. Place the water phase solutions and the oil phase in the wells of a 384-well microtiter plate. The wells must be chosen in correspondence of the script set as described in Subheading3 **step 1**.
6. Add about 2 mm of mineral oil on the top of the microtiter plate to avoid any evaporation and contamination of the solutions.
7. Place the microtiter plate in correspondence of the arm-robot and start the script; from this step the platform will work automatically until the thermocycler module (Subheading3 step10). In detail, the tubing will move from the different wells generating the first train of droplets that contain: (1) the magnetic beads solution, (2) the first total RNA sample, (3) the washing buffer, and (4) RT-qPCR mix. The droplet can have different volumes, however working between 50 and 300 nL is suggested for a better size monodispersion. Between every droplet 500 nL of oil is inspired.
8. The train is then automatically driven in the serpentine of the first PDMS device. Here, the script switch from the state 1 to

the state 2 and the train is driven beyond the chip 2. Switching back to state 1, the script allows the generation of a second train of droplets while the first is pushed by the Syringe 2 forward, between the two couples of magnetic tweezers.

9. When the first droplet containing the magnetic beads passes in correspondence of the first tweezers, the LabView program actives the electric coil and the beads are trapped and extracted. When the next droplet is coming, the program recognizes the meniscus and releases the magnetic beads. This droplet flows for 5 min at 0.1 μL/s until reaching the second magnetic tweezers. During this time the mRNAs bond with the oligo-dT presented on beads, which are then extracted. The resulting cluster is then washed by third droplet and finally dispersed in the last one, which is driven in the thermocycler for the Reverse Transcription.
10. After the RT, the user must take the following operation manually. The droplets containing the resulting cDNA are recovered into 200 μL tubes with 5 μL of DEPC-treated water. These droplets are easy to recognize because they contain the magnetic beads. The solution is then treated as described in paragraph Subheading 2.1 step 5 to obtain the Ct values.
11. Due to the deterministic generation and control of droplets, several combinations of total RNA samples and genes to screen can be performed. Figure 2 shows an example of calibration curves for two genes (TBP and HER2) obtained from total RNA extracted from MCF7 and SKBR3 cell lines. The total RNA extraction protocol is the one developed by Chomczynski and Sacchi [12], based on an acid guanidinium thiocyanate-phenol-chloroform.

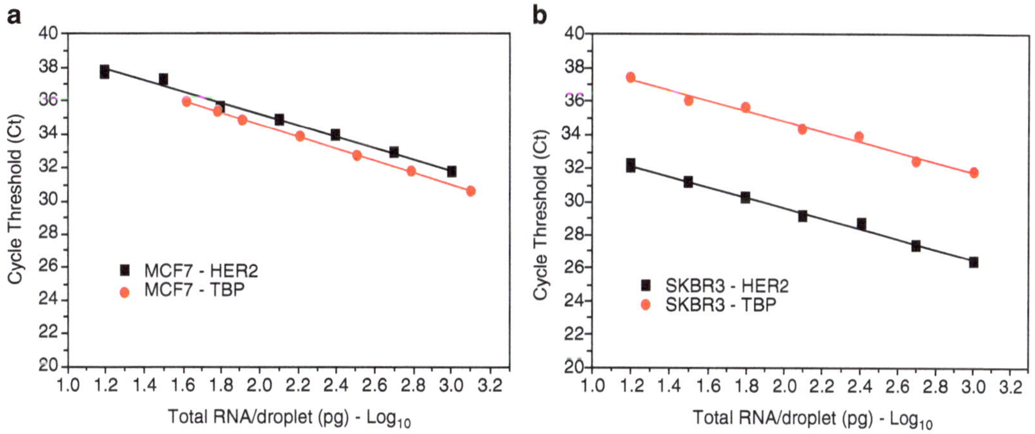

Fig. 2 Calibration curves for (**a**) MCF7 and (**b**) SKBR3 cell lines for genes HER2 and TBP, represented with *black squared* and *red circle* dots respectively: CTs are plotted in logarithmic scale as a function of the total RNA concentration. The qPCR efficiency E*ff*, reported below the fit parameters, was extracted from the slope of the linear fits

4 Notes

1. All the experiment with the microfluidic platform can be performed in a conventional chemical-physics laboratory where any precaution concerning the RNA manipulation is taken into account. In fact, the droplet configuration allows working in a close system, which is never in contact with the external environment once that the droplet is formed.
2. The presence of air bubble in the capillary or in the syringe pumps could generate problems during the droplet generation. In particular, the air bubbles act as dampener in the fluidic system. Therefore, the volume of the water phase droplet could be different than the desired values. Higher is the air volume trapped, higher is this effect.
3. With the presented setup to obtain monodisperse droplet in size (about 2 %) working with volume between 50 nL and 300 nL is recommended. However, it is possible to decrease the size of the droplet keeping a good precision using a smaller syringe (for example, the 100 μL SGE). In this case it can be useful to reduce the capillary size also.
4. The PMDS chips prepared by standard soft-lithography present channels size compatible with the capillary that transports the droplet along the platform. Therefore, if the capillary is changed for example to use larger or smaller droplet volume, the PDMS devices must be adapted. Additionally, the inlets and the outlets capillaries are fixed parallel to the channels to avoid any droplet splitting.

Acknowledgments

This work was supported by the NADINE FP7 project, ANR DigiDiag (ANR 10-NANO-02-03) and by the Foundation ARC (young researcher fellowship D.F.). Some technological elements used in this article are covered by patent WO2013041983, property of Centre National de la Recherche Scientifique.

References

1. Vogelstein B, Kinzler KW (1999) Digital PCR. P Natl Acad Sci USA 96:9236–9241
2. Mary P, Dauphinot L, Bois N et al (2011) Analysis of gene expression at the single-cell level using microdroplet-based microfluidic technology. Biomicrofluidics 5:024109
3. Marcus JS, Anderson WF, Quake SR (2006) Parallel picoliter RT-PCR assays using microfluidics. Anal Chem 78:956–958
4. Eastburn DJ, Sciambi A, Abate AR (2013) Ultrahigh-throughput Mammalian single-cell reverse-transcriptase polymerase chain reaction in microfluidic drops. Anal Chem 85:8016–8021
5. Rival A, Jary D, Delattre C et al (2014) An EWOD-based microfluidic chip for single-cell isolation, mRNA purification and subsequent multiplex qPCR. Lab Chip 14:3739–3749

6. Seemann R, Brinkmann M, Pfohl T, et al (2012) Droplet based microfluidics. Reports on progress in physics. Physical Society (Great Britain) 75: 016601.
7. Ali-Cherif A, Begolo S, Descroix S et al (2012) Programmable magnetic tweezers and droplet microfluidic device for high-throughput nanoliter multi-step assays. Angew Chem Int Ed 51:10765–10769
8. Teste B, Ali-Cherif A, Viovy JL et al (2013) A low cost and high throughput magnetic bead-based immuno-agglutination assay in confined droplets. Lab Chip 13:2344–2349
9. Chabert M, Dorfman KD, de Cremoux P et al (2006) Automated microdroplet platform for sample manipulation and polymerase chain reaction. Anal Chem 78: 7722–7728
10. Guarneri V, Barbieri E, Dieci MV et al (2010) Anti-HER2 neoadjuvant and adjuvant therapies in HER2 positive breast cancer. Cancer Treat Rev 36(Suppl 3): S62–S66
11. Chavez-MacGregor M, Zhang N, Buchholz TA et al (2013) Trastuzumab-related cardiotoxicity among older patients with breast cancer. J Clin Oncol Off J Am Soc Clin Oncol 31:4222–4228
12. Chomczynski P, Sacchi N (1987) Single-step method of RNA isolation by acid guanidinium thiocyanate-phenol-chloroform extraction. Anal Biochem 162:156–159

Chapter 10

SlipChip Device for Digital Nucleic Acid Amplification

Feng Shen

Abstract

Digital nucleic acid amplification (Digital NAA) quantifies nucleic acid by compartmentalizing a sample of DNA or RNA into a large number of discrete partitions and performing parallel nucleic acid amplification, such as polymerase chain reaction (PCR) or isothermal amplification reactions. With the counts of positive wells, total number of wells, and volumes of wells, the concentration of the target nucleic acid in the sample can be quantified. Digital NAA is considered increasingly powerful for ultra-sensitive detection and accurate quantification of nucleic acid for biological research and potentially medical diagnostics. Here, we describe glass SlipChip devices to perform digital NAA without cumbersome manual manipulation or complex fluidic control systems.

Key words Digital PCR, Quantification, Nucleic acid amplification, Microfluidics, Lab on a chip, Ultra-sensitive detection, Single molecule

1 Introduction

Digital nucleic acid amplification (digital NAA) recently becomes an increasingly powerful tool for ultrasensitive detection and accurate quantification of nucleic acid [1–3]. Digital NAA compartmentalizes a sample of DNA or RNA into a large number of partitions, which are often in sub-microliter size. Following the *Poisson* distribution, each partition may either contain the target nucleic acid "1" or contain no nucleic acid of interest "0". The nucleic acid in each individual partition can be efficiently amplified in parallel to reach an end-point, which can be easily distinguished from the no-template wells. By counting the number of positive wells (or negative wells), the concentration of the target nucleic acid can be calculated with the information of total number of wells and volumes of wells. Since digital NAA is less dependent on real-time reaction kinetics, it can potentially provide more repeatable results and potentially be more tolerant to environment perturbation, inhibitors, or competing background reactions [3]. Since digital NAA can amplify single molecule of nucleic acid in a "digital" format, it is often considered a sensitive quantification

Valérie Taly et al. (eds.), *Microchip Diagnostics: Methods and Protocols*, Methods in Molecular Biology, vol. 1547,
DOI 10.1007/978-1-4939-6734-6_10, © Springer Science+Business Media LLC 2017

method that can provide accurate quantification of nucleic acid with finer resolution comparing to real-time nucleic acid amplification, such as real-time qPCR [1–3].

Based on the thermal profiles for amplification reactions, digital NAA can be considered as two major groups: digital PCR, which requires thermal cycling for PCR, and digital isothermal nucleic acid amplification, which only needs constant temperature for reaction. There are several digital PCR systems available for research applications [1]. They can be considered as two major categories: droplet-based digital PCR [1, 4] and chip-based digital PCR [1, 5]. Droplet-based systems can generate a large number of partitions (droplets) in a high-throughput format, but it generally requires instruments for droplet generation and analysis. Furthermore, it is less flexible with selection of amplification reagents since it requires specific surface chemistry to ensure the droplets are stable and are uniform in size, which is critical for accurate digital quantification. For chip-based digital PCR, it either needs cumbersome manual steps for device handling or requires complex fluidic manipulation systems.

SlipChip provides a unique solution to perform digital NAA, including digital PCR [6], digital RT-PCR [7], and digital isothermal nucleic acid amplification [8, 9]. In general, a SlipChip device consists of two plates that have features, such as wells or channels, imprinted on the facing surface of each plate [10]. Partially overlapping the features alternatively can generate continuous fluidic paths. After introducing a sample into the device, one plate can be "slipped" relatively to the other plate to disconnect the fluidic paths and to generate discrete partitions in parallel (Fig. 1). Then, the

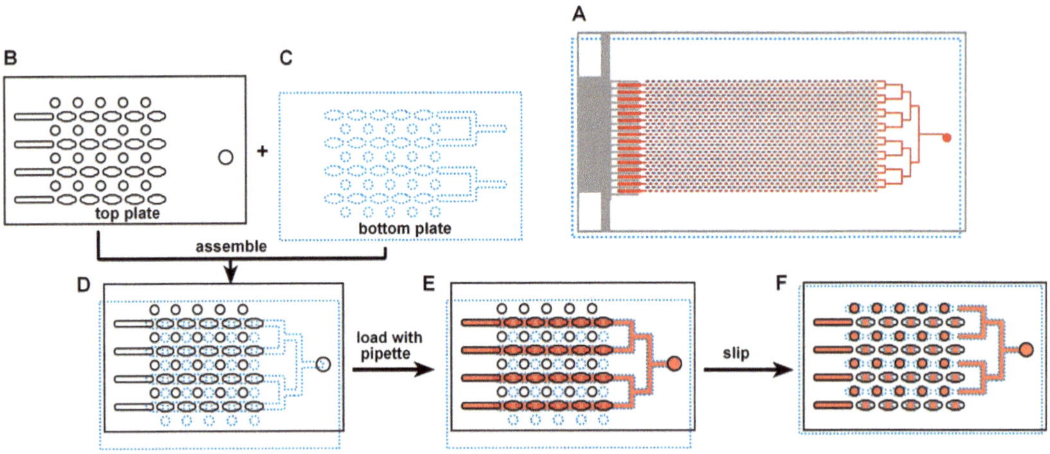

Fig. 1 Schematic drawings indicate assembly and operation of the digital PCR SlipChip. (**a**) Schematic drawing indicates the design of the entire assembled SlipChip for digital amplification. (**b**) Schematic drawing shows the design of top plate. (**c**) Schematic drawing shows the design of bottom plate. (**d**) Schematic drawing shows the assembled SlipChip. (**e**) The reagent (red) can be injected into SlipChip through the continuous fluidic path. (**f**) The bottom plate was slipped relative to the top plate to form individual reaction compartment. Reproduced from [6] with permission from Royal Society of Chemistry)

SlipChip device can be placed on an in situ adaptor in a thermal cycler for nucleic acid amplification. Several samples can be characterized in parallel on the same SlipChip device to increase the throughput [7, 11]. A multivolume SlipChip device was also developed to cover a large dynamic range of concentrations with wells of a series of different volumes [7].

SlipChip platform was also demonstrated for a number of other applications, including multiplex PCR [11], genotyping [12], protein crystallization [13], immunoassay [14, 15], dry sample storage [16], bacterial analysis [17], and isoelectric focusing [18]. Axillary features, such as dead-filling features [19], a pumping lid [20], and mobile phone-based imaging and analysis systems [21], were also developed previously.

Here, we focus on the glass SlipChip device with well of single volume for digital PCR application.

2 Materials

Prepare all the solutions for wet etching with Milli-Q™ water (Millipore, Billerica, MA) and prepare all the solutions for PCR amplification with nuclease-free water. Follow all the safety protocol and waste disposal regulations. All solvents and salts purchased from commercial sources were used as received unless otherwise specified.

2.1 Glass SlipChip Fabrication Components

1. Soda-lime glass plates with chromium and photoresist coating.
2. Amorphous diamond coated drill bits with 0.030 inch cutter diameter.
3. A glass etchant solution: 1 mol/L HF, 0.5 mol/L NH_4F, and 0.75 mol/L HNO_3. The concentration of the components can be adjusted accordingly to alternate the etching speed. Danger: the solution can cause skin burn! (*see* **Note 1**).
4. A chromium etchant solution: 0.6 mol/L $HClO_4$ and 0.365 mol/L $(NH_4)_2Ce(NO_3)_6$.
5. Photoresist remover solution: 0.1 mol/L NaOH in water.
6. Ethanol (200 proof).
7. Acetone.
8. Chloroform.
9. Photomasks with designed features in high resolution.
10. Dichlorodimethylsilane (≥99.5 %) was obtained from Sigma-Aldrich (St. Louis, MO). Toxic if inhaled, and it causes severe skin burns and eye damage. Storage temperature 2–8 °C.
11. Piranha solution: freshly mixed 3:1 volume ratio of sulfuric acid and hydrogen peroxide. Piranha solution causes severe skin burns.

12. A thermal water-shaking bath.
13. Mask aligner.
14. Surface Profilometer.
15. Desiccator.

2.2 PCR Amplification Components

1. All primers were rehydrated with nuclease-free water and stored in −20 °C.
2. PCR master mix (2X).
3. Mineral oil (DNase, RNase, Protease free).
4. DEPC-treated and nuclease-free water.
5. Bovine serum albumin (BSA, 20 mg/mL).
6. Genomic DNA of *Staphylococcus aureus* (American Type Culture Collection).
7. PCR tubes and barrier pipette tips (DNase, RNase free).
8. PCR thermal cycler and flat in situ adapter.

2.3 Optical Instruments for Imaging of SlipChip Device

1. Stereoscope (Leica).
2. Fluorescence microscope (Leica DMI 6000) with a 5X/0.15 NA objective and L5 filter.
3. MetaMorph microscopy automation and image analysis software.

3 Methods

All the procedures were carried out at room temperature unless otherwise specified.

3.1 Wet Etching of the Glass SlipChip Device

1. Soda-lime glass plate with chromium and photoresist coating is aligned with a photo mask, and the uncovered photoresist is exposed using a standard exposure protocol (*see* **Note 2**).
2. The exposed glass plate is immediately immersed in 0.1 M of NaOH solution to remove the area of the exposed photoresist. The processed glass plate is rinsed with Milli-Q™ water and then submerged in a chromium etchant to remove the exposed underlying chromium layer. The glass plate is then thoroughly rinsed with Milli-Q™ water and blow dry with nitrogen gas.
3. The back of the glass plate is protected with water-resistant PVC sealing tape, and the glass plate is immersed in a glass etching solution to etch the exposed glass surface. The wet etching protocol is performed in a shake bath with temperature control (40 °C) to ensure uniform etching of designed features (*see* **Note 1**).

4. The exact depth, roughness, and uniformity of etched features can be characterized using a profilometer.
5. After etching, the glass plate is thoroughly rinsed with Milli-Q™ water to remove the residue etching solution, and then the plate is dried with nitrogen gas.
6. The remaining unexposed photoresist material is removed by ethanol or acetone, and then the plate is immersed in the chromium etchant again to remove the remaining chromium layer.
7. The glass plate is then thoroughly rinsed with Milli-Q™ water to remove the residue chromium etchant, and the plate is dried with nitrogen gas.

   Through holes are drilled in the top plate as the fluidic inlets.

3.2 Surface Treatment of the Glass SlipChip Device

1. The glass plate is dried with nitrogen gas and then placed in a plasma cleaner to oxidize the glass surface for approximately 100 s.
2. The oxidized glass plate is immediately transferred into a desiccator with 50 μL of dichlorodimethylsilane (*see* **Notes 3** and **4**).
3. A vacuum is applied to provide negative pressure in the desiccator for vapor phase silanization for approximately 1 h.
4. The glass plate is taken out from the desiccator and thoroughly rinsed with chloroform, acetone, and ethanol sequentially to remove the remaining silane (*see* **Note 5**).
5. The glass plate is thoroughly dried with nitrogen gas.

3.3 Assembly of the SlipChip Device

1. The bottom plate is immersed into mineral oil with the patterned side facing up (*see* **Note 6**).
2. The top plate is placed on the top of the bottom plate with the pattern features aligned with the features on the bottom plate to establish continuous fluidic paths (Fig. 1a–d) (*see* **Note 7**).
3. The two plates are then fixed using binder or clamping fixtures (*see* **Note 8**).

3.4 Operation of the SlipChip Device

1. The reaction mixture contains 10 μL of 2X SsoFast EvaGreen SuperMix, 2 μL of 10 mg/mL BSA solution, 0.5 μL of forward primer (10 μM), 0.5 μL of reverse primer (10 μM), 2 μL of *S. aureus* genomic DNA template (replaced with 2 μL water in no template negative control experiments), and 5 μL of nuclease-free water.
2. This reaction mixture is introduced into the SlipChip device by direct pipette loading (Fig. 1d–e).
3. After the solution is completely loaded into the device, the top plate is slipped relatively to the bottom plate to break the fluidic paths and to overlay the wells on one plate with thermal expansion well on the facing plate (Fig. 1e–f).

4. Multiple template solution can be tested on the same SlipChip device by introducing through different loading ports.

3.5 Performing Digital PCR on the SlipChip Device

1. The SlipChip device is placed on an in situ adaptor in the thermal cycler for thermal cycling.
2. To amplify *S. aureus* genomic template, the following protocol shall be performed: an initiation step of 2 min at 94 °C; A total of 35 cycles of amplification with 1 min at 94 °C, 30 s at 55 °C, and 30 s at 72 °C; after the final cycle, a final extension step of 5 min at 72 °C (*see* **Note 9**).
3. The SlipChip device is allowed to cool down to room temperature before imaging and further analysis.

3.6 Imaging of the SlipChip Device

1. Bright field images are acquired by stereoscope for characterization of alignment accuracy.
2. Fluorescence images before and after thermal cycling are acquired by using epi-fluorescence microscope with a 5X/0.15 NA objective and L5 filter (*see* **Note 10**).
3. The threshold level of all fluorescence images is adjusted to the same value.
4. All fluorescence images are background-corrected by an image obtained with a standard fluorescence slide.
5. The wells loaded with template have significant increase of the fluorescence intensity compared to the no template control wells. A line scan across the wells can be applied for analysis of fluorescence intensity (Fig. 2).
6. Amplification product can be recovered after thermal cycling for further analysis, such as gel-electrophoresis to characterize the molecular weight of amplicons (*see* **Note 11**).

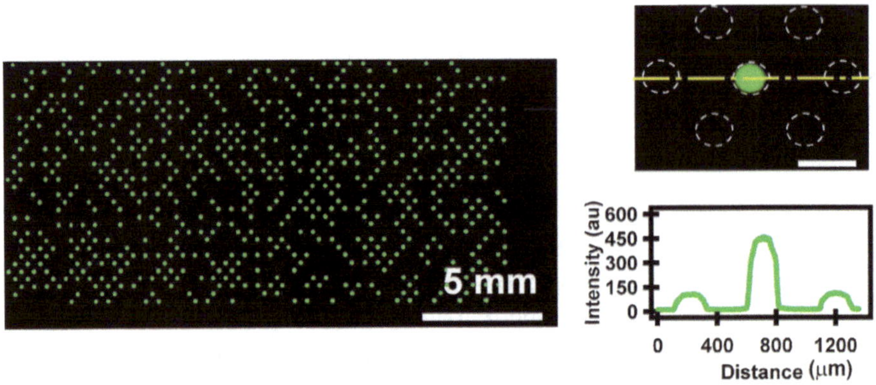

Fig. 2 Fluorescence image and line scan at the end-point of amplification on the digital PCR SlipChip. Reproduced from [6] with permission from Royal Society of Chemistry)

Fig. 3 Bright field image of multivolume digital SlipChip loaded with food dye (*left*); Fluorescent photograph (*middle* and *right*) of a multiplexed digital RT-PCR detection panel: (I) measurement of internal control RNA template; (II) HCV control viral RNA quantification; (III) negative control for HIV; (IV) HIV viral RNA quantification; (V) negative control for HCV. Inset shows an amplified area from HCV viral load test. Reproduced from [7] with permission from American Chemical Society

3.7 Data Analysis for Digital PCR on the Single Volume SlipChip Device

A regression fit of linearized form of the *Poisson* equation (Eq.1) is applied to characterize the performance of digital PCR SlipChip [6]:

$$\ln\left(n-f\right) = -c \times x + \ln n \tag{1}$$

In Eq. (1): n is the total number of wells; f represents the number of positive wells; c is the initial concentration of the sample; x is the fractional dilution factor of the original concentration. The total number of empty wells is $n - f$. From this equation, plotting x vs. $\ln(n - f)$ with constrain of the intercept to $\ln(n)$ can provide a slope that equals the negative of the original concentration.

3.8 Multivolume Digital SlipChip for Quantification Over Large Dynamic Range

Multivolume digital SlipChip can achieve a large dynamic range with a much lower total number of wells compared to single volume platform (Fig. 3). It can provide flexibility to adjust dynamic range and resolution separately depending on specific applications [22]. The smallest wells provide quantification of template at high concentration, and the large-volume wells increase the lower detection limit by efficiently increasing the total reaction volume. At high concentration, all the wells of large volumes will contain more than one copy of template, and digital amplification will be achieved at wells of small volumes. At low concentration, all the wells of small volumes will be statistically "empty," and digital quantification will be achieved at wells of large volumes.

3.9 Digital Isothermal Amplification on the SlipChip Device

Digital isothermal NAA has great potential for quantitative molecular diagnostics in resource-limited settings. However, it is nontrivial to achieve accurate quantification using digital isothermal NAA. Unlike "hot start" PCR polymerase, which is activated by high reaction temperature, isothermal amplification enzymes can have "background" pre-amplification at temperatures close to room temperature. This pre-amplification can take place as soon as all the reaction components are mixed together and also during generation of reaction partitions for digital amplification [8]. As a consequence, this approach may provide a higher concentration than its true value. For example, recombinase polymerase amplification (RPA) is designed and optimized for isothermal amplification at 39 °C, but it can still amplify template at a slow background rate at room temperature (~25 °C). To minimize the background pre-amplification, RPA is recommended to add magnesium acetate ($Mg(OAc)_2$) solution as the last component, which serves as a chemical initiator for the amplification reaction.

Performing digital isothermal NAA with chemical initiation requires fluidic handling of a large number of reaction partitions. Droplet fusion has been demonstrated to merge two droplets for digital isothermal amplification before [23]; however, this method requires reliable pair formation and precise fluidic control. The digital SlipChip device provides a simple and robust workflow, which compartmentalizes the sample with template first, then delivers the chemical initiator to each compartment in parallel with additional slipping [8] (Fig. 4).

4 Notes

1. Glass etchant is highly corrosive, wearing gloves and personal protection equipment. In case of accident or exposure, apply Calgonate® (Port St Lucie, FL) and seek aid immediately.
2. Actual exposure time needs to be characterized and adjusted for individual light source.
3. Dichlorodimethylsilane reacts violently with water, and it needs to be stored sealed under 2–8 °C.
4. Dichlorodimethylsilane is corrosive to the respiratory tract. Experiments need to be performed in a fume hood with proper personal protection equipment.
5. Flammable solvent. Cleaning process needs to be performed in a fume hood.
6. Device needs to be assembly under clean environment to avoid contamination, for instance nuclease contamination or template/amplicons contamination.
7. Alignment of the two plates can be performed under stereoscope for better accuracy.
8. Binder clips can be used to fix the two plates of SlipChip device.
9. Annealing temperature needs to be adjusted according to the primer design.

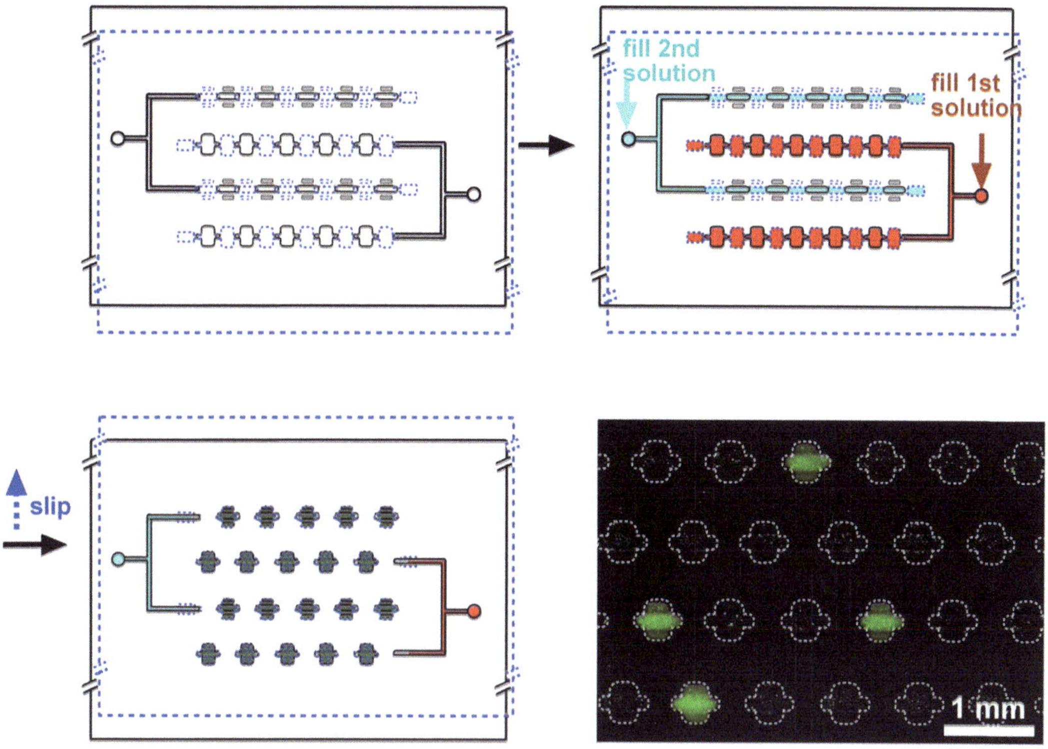

Fig. 4 Schematic drawing indicates the workflow of partitioning and combining 1st and 2nd solutions in a SlipChip device. Fluorescence image indicates digital RPA using parallel delivery of chemical initiator to each partition. Reproduced from [8] with permission from American Chemical Society

10. Exposure time needs to be adjusted according to the intensity of light source. Overexposure may cause photo bleaching of fluorescence dye or probe.

11. Recover amplification product from SlipChip device or open the device after amplification may increase the risk of contamination of working environment. The devices need to be thoroughly cleaned with piranha solution. Piranha solution is highly corrosive, wearing proper personal protection equipment.

References

1. Baker M (2012) Digital PCR hits its stride. Nat Methods 9(6):541–544
2. Huggett JF et al (2013) The digital MIQE guidelines: minimum information for publication of quantitative digital PCR experiments. Clin Chem 59(6):892–902
3. Witters D et al (2014) Digital biology and chemistry. Lab Chip 14(17):3225–3232
4. Hindson BJ et al (2011) High-throughput droplet digital PCR system for absolute quantitation of DNA copy number. Anal Chem 83(22):8604–8610
5. Ottesen EA, Hong JW, Quake SR, Leadbetter JR (2006) Microfluidic digital PCR enables multigene analysis of individual environmental bacteria. Science 314(5804):1464–1467
6. Shen F, Du W, Kreutz JE, Fok A, Ismagilov RF (2010) Digital PCR on a SlipChip. Lab Chip 10(20):2666–2672
7. Shen F et al (2011) Multiplexed quantification of nucleic acids with large dynamic range using multivolume digital RT-PCR on a rotational SlipChip tested with HIV and hepatitis C viral load. J Am Chem Soc 133(44):17705–17712

8. Shen F et al (2011) Digital isothermal quantification of nucleic acids via simultaneous chemical initiation of recombinase polymerase amplification reactions on SlipChip. Anal Chem 83(9):3533–3540
9. Sun B et al (2013) Mechanistic evaluation of the pros and cons of digital RT-LAMP for HIV-1 viral load quantification on a microfluidic device and improved efficiency via a two-step digital protocol. Anal Chem 85(3):1540–1546
10. Du W, Li L, Nichols KP, Ismagilov RF (2009) SlipChip. Lab Chip 9(16):2286–2292
11. Shen F et al (2010) Nanoliter multiplex pcr arrays on a slipchip. Anal Chem 82(11):4606–4612
12. Sun B et al (2014) Measuring fate and rate of single-molecule competition of amplification and restriction digestion, and its use for rapid genotyping tested with hepatitis C viral RNA. Angew Chemie Int Ed 53(31):8088–8092
13. Li L, Du W, Ismagilov R (2010) Multiparameter screening on slipchip used for nanoliter protein crystallization combining free interface diffusion and microbatch methods. J Am Chem Soc 132(1):112–119
14. Liu W, Chen D, Du W, Nichols KP, Ismagilov RF (2010) SlipChip for immunoassays in nanoliter volumes. Anal Chem 82(8):3276–3282
15. Ge S, Liu W, Schlappi T, Ismagilov RF (2014) Digital, ultrasensitive, end-point protein measurements with large dynamic range via brownian trapping with drift. J Am Chem Soc 136(42):14662–14665
16. Begolo S, Shen F, Ismagilov RF (2013) A microfluidic device for dry sample preservation in remote settings. Lab Chip 13(22):4331–4342
17. Ma L et al (2014) Gene-targeted microfluidic cultivation validated by isolation of a gut bacterium listed in human microbiome project's most wanted taxa. Proc Natl Acad Sci U S A 111(27):9768–9773
18. Zhao Y, Pereira F, deMello AJ, Morgan H, Niu X (2014) Droplet-based in situ compartmentalization of chemically separated components after isoelectric focusing in a Slipchip. Lab Chip 14(3):555–561
19. Li L, Karymov MA, Nichols KP, Ismagilov RF (2010) Dead-end filling of slipchip evaluated theoretically and experimentally as a function of the surface chemistry and the gap size between the plates for lubricated and dry slipchips. Langmuir 26(14):12465–12471
20. Begolo S, Zhukov DV, Selck DA, Li L, Ismagilov RF (2014) The pumping lid: investigating multi-material 3D printing for equipment-free, programmable generation of positive and negative pressures for microfluidic applications. Lab Chip 14(24):4616–4628
21. Selck DA, Karymov MA, Sun B, Ismagilov RF (2013) Increased robustness of single-molecule counting with microfluidics, digital isothermal amplification, and a mobile phone versus real-time kinetic measurements. Anal Chem 85(22):11129–11136
22. Kreutz JE et al (2011) Theoretical design and analysis of multivolume digital assays with wide dynamic range validated experimentally with microfluidic digital PCR. Anal Chem 83(21):8158–8168
23. Mazutis L et al (2009) Droplet-based microfluidic systems for high-throughput single DNA molecule isothermal amplification and analysis. Anal Chem 81(12):4813–4821

Chapter 11

Multiplex Detection of *KRAS* Mutations Using Passive Droplet Fusion

Deniz Pekin and Valerie Taly

Abstract

We describe a droplet microfluidics method to screen for multiple mutations of a same oncogene in a single experiment using passive droplet fusion. Genomic DNA from H1573 cell-line was screened for the presence of the six common mutations of the *KRAS* oncogene as well as wild-type sequences with a detection efficiency of 98 %. Furthermore, the mutant allelic fraction of the cell-line was also assessed correctly showing that the technique is quantitative

Key words Droplet microfluidics, Digital PCR, Multiplexing, Passive droplet fusion, Biomarkers, *KRAS*, Cancer.

1 Introduction

Multiplexing is a term used to describe the simultaneous testing for multiple mutations. In clinical analyses, often a very limited knowledge on the mutational status of the original tumor is available. Especially for biomarkers such as *KRAS* where several mutations exist, a multiplex screening enabling the simultaneous detection of multiple mutations is required. Multiplexing also minimizes the time and sample consumption.

With their high-throughput nature, droplet-based microfluidic technologies are pertinent tools to run several assays in parallel. In order to realize such multiplexing in droplets, a fluorescent labeling is necessary for the discrimination of the different assays or components during the assay readout. The "one target = one fluorophore" strategy allows an easy discrimination of populations but it cannot be used on a large extent while keeping a reasonable number of emission channels, especially in applications such as testing for the mutational status of the *KRAS* gene where seven common mutations exist.

Here, we describe a strategy that uses increasing concentrations of a dye to code for different components inside droplets and the

Valérie Taly et al. (eds.), *Microchip Diagnostics: Methods and Protocols*, Methods in Molecular Biology, vol. 1547, DOI 10.1007/978-1-4939-6734-6_11, © Springer Science+Business Media LLC 2017

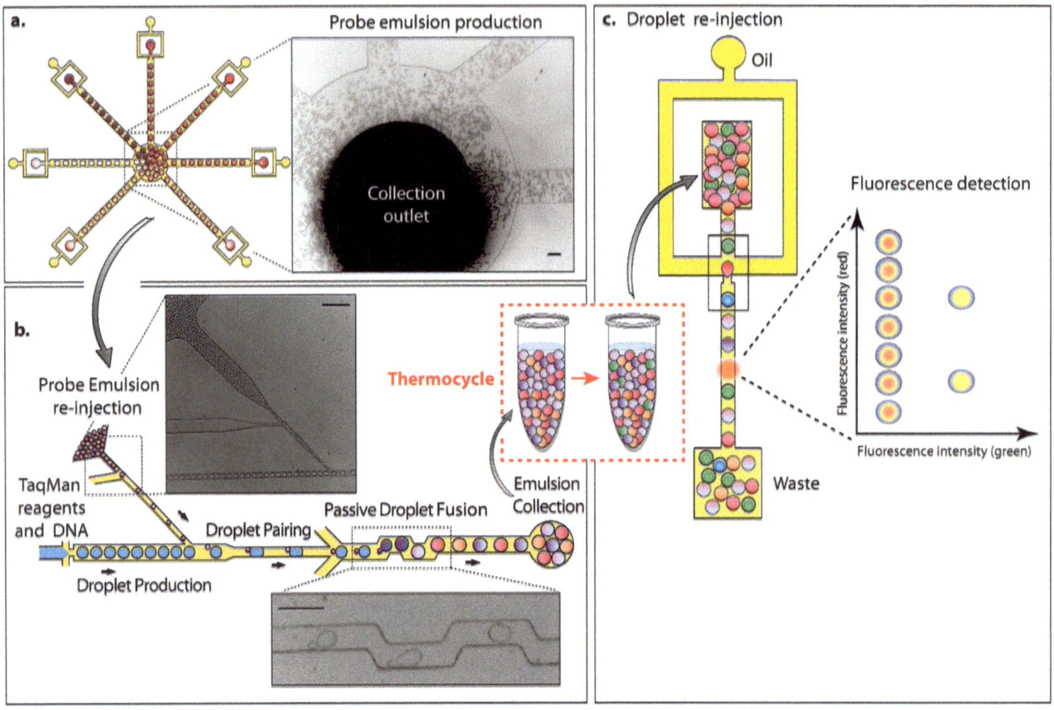

Fig. 1 The procedure of the multiplexing PCR using passive droplet fusion: (**a**) the probe emulsion is produced using a multiple droplet maker device and collected off-chip. (**b**) the probe emulsion is reinjected on a passive fusion device and each droplet is fused with on-chip droplets containing genomic DNA. The obtained emulsion is collected off-chip and thermocycled (*inner panel*). The thermocycled emulsion is reinjected for fluorescence analysis. The sketched histogram represents droplet clusters appearing according to the Dextran Texas Red code and the expected green fluorescent droplet clusters. (scale bars: 100 μm)

recent advances in microfluidic droplet manipulation technology such as the passive droplet fusion [1] to screen for the common *KRAS* mutations of the 12th codon in a single experiment (see [2]) (Fig. 1).

2 Materials

2.1 Cell Lines

SW48 (American Type Culture Collection ATCC CCL-231, wild-type *KRAS*).

H1573 (ATCC CRL-5877, transversion G to C in the second base of *KRAS* codon 12, G12 A).

2.2 Genomic DNA Preparation (See Notes 1–4)

QIAamp DNA mini Kit (Qiagen), following manufacturer's protocol.

Sterile nuclease-free water.

Nanodrop ND-1000 spectrophotometer (Thermo Scientific, USA) for absorbance measurements.

2.3 Materials for Droplet Microfluidics

PHD 2000 (Harvard Apparatus, France) syringe pumps.

A home-made Labview (National Instruments) interface [3, 4].

1 mL polyethylene syringes (Omnifix-FR, BRAUN, Germany).

23G syringe needles.

0.56 mm × 1.07 mm polytetrafluoroethylene (PTFE) tubings.

HFE-7500 oil (3 M) with 2 % (w/w) EA-surfactant (Raindance Technologies, USA).

2.4 Microfluidics Device Preparation and Operation

AutoCAD (Autodesk).

Photolithography Mask (Selba Corp, Switzerland).

PDMS and curing agent (Sylgard® 184 silicone elastomer kit, Dow Corning).

SU8 photoresist and developer (MicroChem Corp).

Silicon wafers (Siltronix).

MJB3 contact mask aligner (SUSS MicroTech).

0.75 mm Uni Core biopsy puncher (Ted Pella Inc., USA).

Oxygen plasma cleaner (Diener Electronic, Germany).

2 % perfluorooctyldimethylsilane (Sigma-Aldrich) in HFE-7100 oil (3 M, USA).

2.5 Microfluidics Device Preparation and Operation (See Notes 3, 6, and 12)

On-chip droplet analysis and fluorescence detection is performed on a microfluidic station as described before [2, 3, 4] (see Fig. 3 in Chap. 12).

2.6 TaqMan Probes and Primers

The TaqMan® MGB probes (Applied Biosystems, Life Technologies) and the primers allowing amplification of the *KRAS* target sequences (Eurogentec) were previously developed and validated for clinical sample analysis [5] (*see* **Notes 3, 6,** and 7). The probes specific for mutated sequences are conjugated to the fluorophore 6-FAM (λex 494 nm/λem 522 nm), whereas the probes complementary to wild-type sequence are conjugated to the fluorophore NED (λex 546 nm/λem 575 nm).

2.7 PCR Reagents (see Notes 2, 7, and 8)

Detergent-free buffer (Finnzymes, Finland) (*see* **Note 9**).

Dextran Texas Red (DTR) dye (Life Technologies).

dNTPs mix (MP Biomedicals).

DyNAzyme II polymerase (5 U/μL, Finnzymes, Finland) (*see* **Note 10**).

2.8 The Collection–Reinjection Device (see Notes 3 and 5)

0.5 mL PCR tubes (Thermo Fisher Scientific).

PDMS plug (created with a 8 mm diameter biopsy punch from a 6 mm thick PDMS disk pre-poured in a Petri dish).

3 Methods

3.1 Genomic DNA Preparation (see Notes 1, 2, 6, and 7)

Extract Genomic DNA (gDNA) from cultured cells using a QIAamp DNA mini Kit following manufacturer's protocol.

Elute gDNA in sterile nuclease-free water (*see* **Note 7**).

Determine purified DNA concentration by measuring the absorbance at 260 nm using a Nanodrop ND-1000 spectrophotometer.

3.2 Microfluidics Device Operation

Microfluidic structures were designed using AutoCAD, a computer-assisted drawing program, and were sent to Selba Tech to be printed as mirrored negative photomasks at a resolution of 25,400 dpi.

Once the photolithography mask is obtained pattern the channels in poly(dimethylsiloxane) (PDMS) using conventional soft lithography techniques [6].

According to the device depth spin coat SU8 photoresist onto silicon wafers and pattern by UV exposure using a MJB3 contact mask aliner through the photolithography mask.

After the exposure spin coat the silicon wafer with the subsequent SU8 developer.

Prepare the PDMS and curing agent as a mixture of 10:1 (w/w).

Degas the mixture under vacuum and pour onto the silicon wafer and place the wafer in a 65 °C incubator for 2 hours for curing.

Peel the cured PDMS from the master and produce input/output ports by punching with a 0.75 mm Uni Core biopsy punch.

Bond the structured side of the PDMS onto a 1 mm thick glass slide using an oxygen plasma.

To decrease the wettability of the aqueous phase on the channel walls inject 2 % perfluorooctyldimethylsilane in HFE-7100 oil. Flush the excess fluorosilane out with Nitrogen.

3.3 The Collection–Reinjection Device

Cut a PDMS plug using a 8 mm biopsy puncher.

Create two holes on the plug with a 1 mm biopsy punch (Fig. 2).

Insert the PDMS plug inside the 0.5 mL PCR tube. The position of the plug inside the tube is determined so that all the emulsion plunges inside the thermocycler well.

Fill the device entirely with oil taking care not to leave any air bubbles inside (*see* **Note 11**).

Use 0.56 mm × 1.07 mm PTFE tubings to ensure proper emulsion collection efficient reinjection.

3.4 The TaqMan® Probe Emulsion (See Note 6)

Prepare each aqueous phase comprising detergent-free buffer, a TaqMan® MGB probe (800 μM), labeled with 6-FAM, targeting one of the six common mutations in codon 12 of the *KRAS*

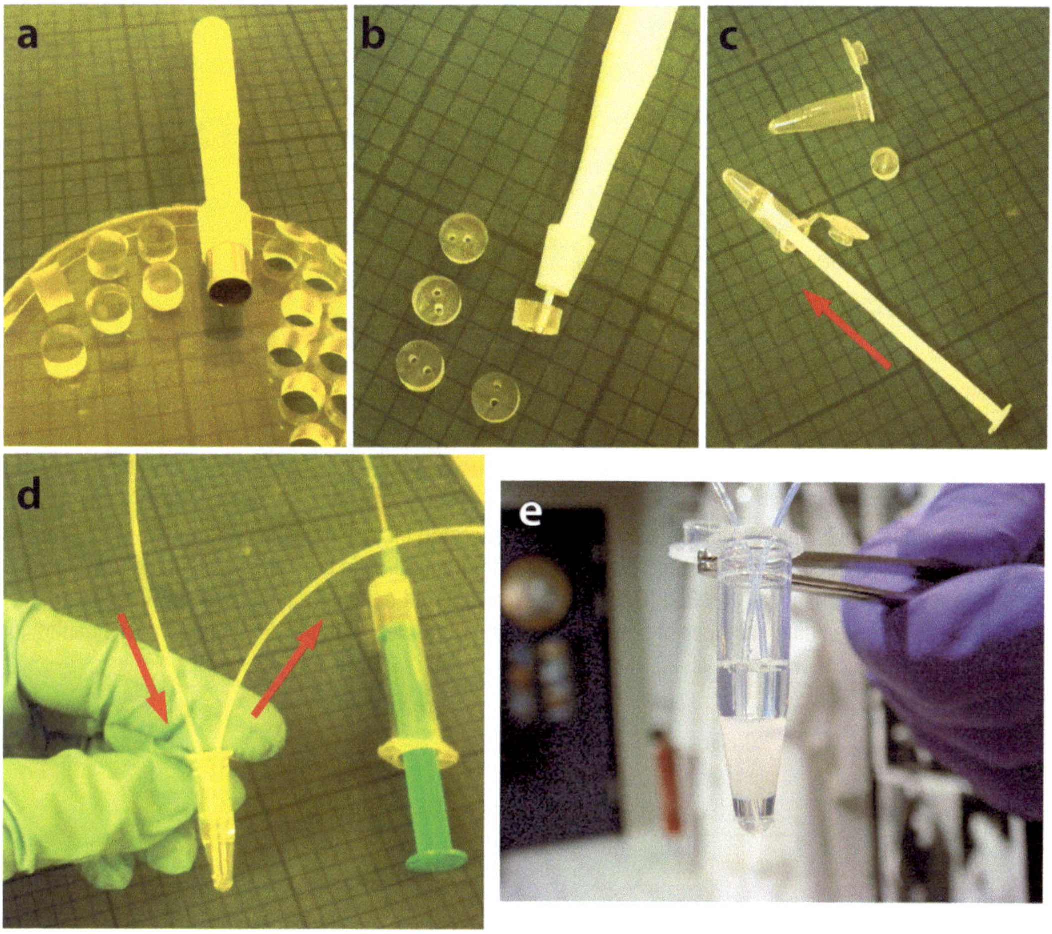

Fig. 2 Collection–reinjection device: A common 8 mm biopsy needle is used to punch plugs from previously prepared PDMS, cured 10:1 (**a**) and a 1 mm biopsy needle is used to punch two holes in the plug for the in/out tubings (**b**). The plug is then inserted into a 0.5 mL PCR tube by pushing with a syringe plunger (**c**). A tubing, connected to a syringe filled with oil, is inserted from one of the holes on the PDMS plug to the bottom of the tube and a second tubing is inserted from the other hole to the end of the PDMS plug (**d**). The vial is filled with oil from the bottom to the top. The second tubing helps the air bubble to come out. It is important at this step to not leave any air inside the tube since the expansion of the air during thermocycle will press the droplets against each other and cause coalescence. (**e**) Collection–reinjection device filled with emulsion

oncogene, or the wild-type *KRAS* codon 12, and a specific concentration of the Dextran Texas Red (DTR), dye in a separate syringe (Fig. 3).

Use a multiple droplet maker device (Fig. 3) to simultaneously produce seven different types of 3 pL droplets (the "probe emulsion") by flow-focusing the aqueous phases with HFE-7500 oil containing 2 % (w/w) EA-surfactant.

Collect the emulsion off-chip in a syringe.

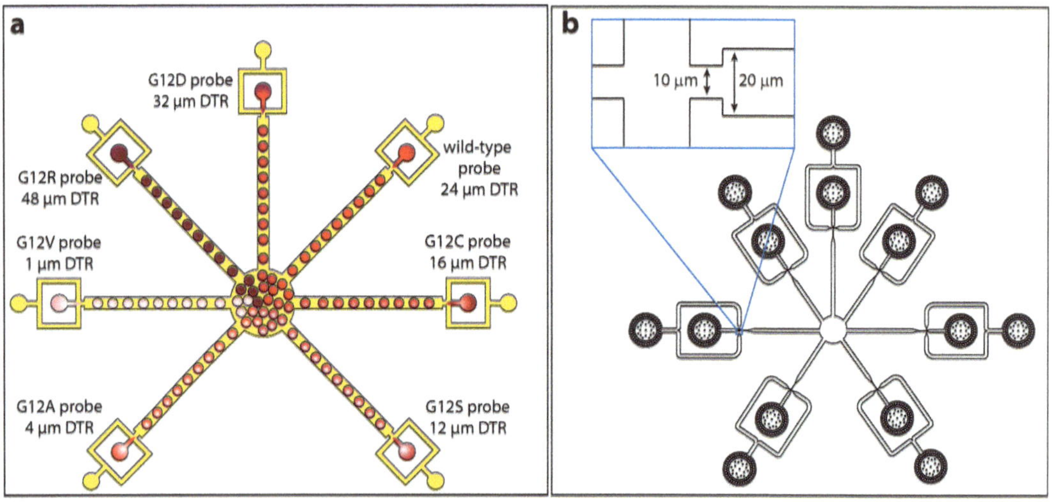

Fig. 3 The probe emulsion: (**a**) a probe emulsion was generated including seven different types of droplets. Each type of droplet contained a TaqMan® probe specific for a mutation in the 12th codon of the *KRAS* gene or wild-type *KRAS*, and Dextran Texas Red (DTR) dye. Seven different concentrations of the DTR dye were used to distinguish the seven different types of droplets. (**b**) the AutoCAD design of the device used for the generation of the "probe emulsion" with a 10 vs. 20 μm nozzle and a depth of 20 μm

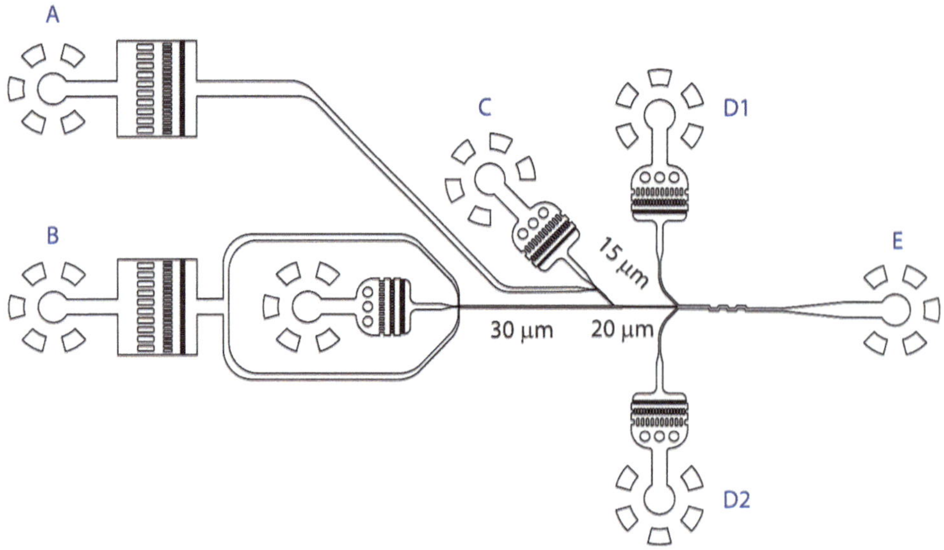

Fig. 4 Passive fusion device:(**a**) Inlet for HFE-7500 oil containing 2 % surfactant. (**b**) Inlet for HFE-7500 oil containing 0.5 % surfactant. (**c**): Inlet for the reinjected emulsion. (**d1** and **d2**): Inlets for HFE-7500 oil containing 2 % surfactant. (**e**): The outlet for droplet collection

3.5 Passive Droplet Fusion

Reinject the probe emulsion into a passive fusion device (Fig. 4) where the droplets will be spaced by HFE-7500 oil and fused with 9 pL droplets produced on-chip (at 1.7 kHz) with HFE-7500 oil containing 0.5 % (w/w) EA surfactant.

The 9 pL droplets contain the target DNA; detergent-free buffer, 200 μM dNTPs, 0.3 μM forward primer, 0.3 μM reverse primer, and 5 units of DyNAzyme II polymerase (5 U/μL). 2 μg of gDNA from the cell line H1573 was used as target DNA.

Collect the generated emulsion inside the "collection/reinjection" device.

3.6 Thermal Cycle

After the removal of the tubings, the device containing the emulsion was placed in a PCR Machine (Mastercycler®, Eppendorf, Le Pecq, France) and the emulsion was thermocycled, starting with an initial denaturation step of 2 min at 94 °C, followed by 39 cycles of: 94 °C, 15 s; 60 °C, 1 min.

3.7 Results

This method was developed for the simultaneous detection and quantification of the six mutations that can affect the 12th codon of the *KRAS* gene as well as the wild-type sequence in a single experiment. The method comprised three steps:

The production of a "probe emulsion" containing seven different types of 3 pL droplets. Each type of droplets contained a TaqMan® probe, labeled with 6-FAM, targeting the six most common mutations in codon 12 of the *KRAS* gene or the wild-type KRAS codon 12. In addition, each type of droplets contained a different concentration of the red fluorescent DTR dye, which served as a code to identify the *KRAS* sequence targeted by the probe in each droplet.

Each on-chip produced 9 pL droplets were fused with a single 3 pL droplet from the reinjected probe emulsion.

The emulsion obtained after fusion was thermocycled and the fluorescence of each droplet was measured.

As a proof of principle for this strategy, we analyzed 12 μg of gDNA from the heterozygous cell-line H1573, which contains a nearly 1:1 ratio KRAS genes with a G to C transversion in the second base of codon 12 and wild-type KRAS genes. Using a 7:2 ratio of 3 pL:9 pL droplets, 60 % of 9 pL droplets were fused with a single 3 pL droplet. Figure 5 shows the results obtained with this strategy. Plotting red fluorescence versus droplet width allowed droplets containing the seven different probes to be identified and droplets arising from uncontrolled coalescence to be excluded from subsequent analysis (Fig. 5). The apparent droplet width increases with increasing red fluorescence [7]; therefore, separate width gates were defined for each population. The red and green fluorescence of the gated droplets was then plotted (Fig. 5). Two populations of green-fluorescent droplets were identified, corresponding to droplets containing TaqMan® probes for wild-type codon 12 and the G to C transversion in the second base of codon 12: 7.6 % of droplets with the wild-type code were green-fluorescent, 8.0 % of droplets with the code for the G to C transversion of codon 12 were green-fluorescent. This corresponds to a mutant

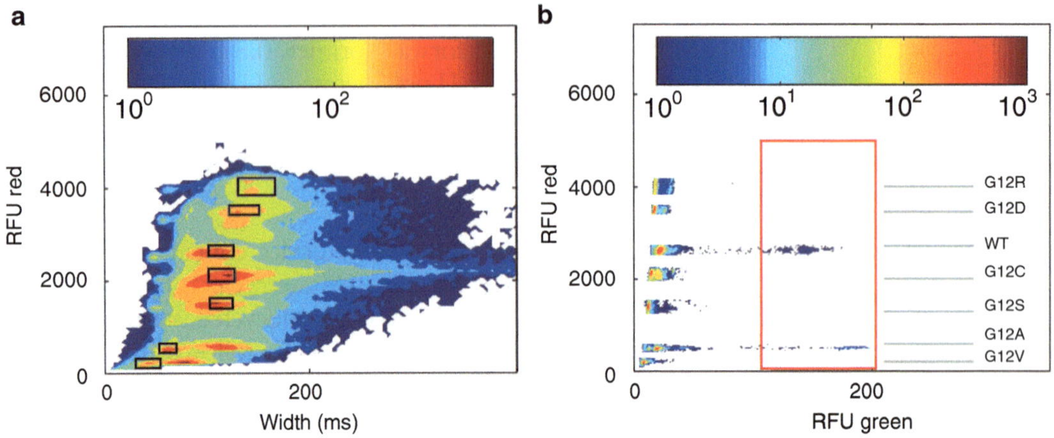

Fig. 5 Fluorescence analysis: The thermocycled emulsion was reinjected for fluorescence analysis. For each droplet, in the red channel, the peak width (droplet size) and maximum peak fluorescence (DTR code) were recorded, and in the green channel, the maximum peak fluorescence (TaqMan® probe signal) was recorded (**a**). The seven droplets containing the probes targeting six different mutations of the *KRAS* gene (121GC, 122GA, 121GA, 122GC, 122GT) and the probe targeting the wild-type *KRAS* (WT) were identified. Panel (**b**) shows a histogram where red fluorescent signal is plotted versus the green fluorescent signal. The probe targeting the 122GC mutation (carried in the H1573 cell-line) and the probe targeting the wild-type sequences gave a green fluorescent signal as expected. The emergence of the additional populations was concluded to be due to uncontrolled droplet coalescence and needed to be eliminated from the analysis. The graphs contain 50,000 bins (500 for the RFU red axis and 100 for the width axis) (Figure modified from [2], permission from Lab on Chip)

allelic fraction (mA%) of 51 %, very close to the expected value of 49.3 % [2]. As expected, 1.75 % of all other populations contained green fluorescent droplets.

4 Notes

1. To minimize contamination by exogenous DNA, all DNA manipulations prior to amplification should be performed in a UV-treated environment (Biocap DNA/RNA hood, Fisher Bioblock Scientific, Illkirch, France).
2. Carry out all the experiments using certified DNA-free, sterilized, filtered pipette tips.
3. All disposables (tweezers, PCR tubes, collection–reinjection devices, tubing, etc.) should be UV-irradiated before use.
4. At the final step of the gDNA extraction, the DNA should be eluted in sterile DNase-, RNase-free, DEPC-treated water (Roth, Karlsruhe, Germany) and not the conventional Tris–EDTA Buffer. Being viscous, EDTA may cause a jetting at the droplet production and hinder the emulsion stability.

5. In addition, we recommend using disposable tweezers (Roth, Karlsruhe, Germany) for microfluidic experiments.
6. The TaqMan® probes should be aliquoted upon arrival and stored at 20 °C. A new aliquot should be used for each set of experiments.
7. Dilutions of PCR reagents and dyes should be performed in sterile DNase-, RNase-free, DEPC-treated water.
8. Adding 1 mg/mL Bovine Serum Albumine (BSA) may help in limiting droplet coalescence.
9. Detergents in the aqueous phase have a negative effect on droplet stability. Therefore, a detergent-free PCR buffer should be used.
10. The solution in which most polymerases are provided has a negative effect on emulsion stability. Hence, a highly concentrated enzyme solution (5 U/μL) should be preferred.
11. Make sure there are no air bubbles remaining inside the collection–reinjection device. During the thermocycle the air bubbles will cause the emulsion to coalesce.
12. A new microfluidic device should be used for each set of experiments to limit carry-over effects.

Acknowledgments

This work was supported by Région Alsace, the Ministère de l'Enseignement Supérieur et de la Recherche, the Université de Strasbourg, the Université Paris Descartes, the Centre National de la Recherche Scientifique (CNRS), the Institut National de la Santé et de la Recherche Médicale (INSERM), the Institut National du Cancer (INCa, no. 2009–1-RT-03-US-1), and the Association pour la recherche sur le Cancer (ARC, no. SL220100601375).

References

1. Mazutis L, Baret JC, Griffiths AD (2009) A fast and efficient microfluidic system for highly selective one to one droplet fusion. Lab Chip 9(18):2665–2672
2. Pekin D, Skhiri Y, Baret JC, Le Corre D, Mazutis L, Ben Salem C, Millot F, El Harrak A, Hutchison JB, Larson JW, Link DR, Laurent-Puig P, Griffiths AD, Taly V (2011) Quantitative and sensitive detection of rare mutations using droplet-based microfluidics. Lab Chip 11(13):2156–2166
3. Baret JC, Miller OJ, Taly V, Ryckelynck M, El-Harrak A, Frenz L, Rick C, Samuels ML, Hutchison JB, Agresti JJ, Link DR, Weitz DA, Griffiths AD (2009) Fluorescence-activated droplet sorting (FADS): efficient microfluidic cell sorting based on enzymatic activity. Lab Chip 9(13):1850–1858
4. Mazutis, L, BaretJC, Treacy, P, Skhiri, Y, Araghi, AF, Ryckelynck, M, Taly, V, and Griffiths, A. D. (2009) Multi-step microfluidic droplet processing: kinetic analysis of an in vitro translated enzyme. Lab Chip 9(20), 2902–2908
5. Lievre A, Bachet JB, Le Corre D, Boige V, Landi B, Emile JF, Cote JF, Tomasic G, Penna C, Ducreux M, Rougier P, Penault-Llorca F, Laurent-Puig P (2006) KRAS mutation status is predictive of response to cetuximab therapy

in colorectal cancer. Cancer Res 66(8): 3992–3995

6. Xia YN, Whitesides GM (2006) Soft lithography. Angew Chem Int Ed 37(5):551–575
7. Clausell-Tormos J, Lieber D, Baret JC, El-Harrak A, Miller OJ, Frenz L, Blouwolff J, Humphry KJ, Koster S, Duan H, Holtze C, Weitz DA, Griffiths AD, Merten CA (2008) Droplet-based microfluidic platforms for the encapsulation and screening of Mammalian cells and multicellular organisms. Chem Biol 15(5):427–437

Chapter 12

Droplet-Based Microfluidics Digital PCR for the Detection of KRAS Mutations

Deniz Pekin and Valerie Taly

Abstract

We demonstrate an accurate and sensitive quantification of mutated *KRAS* oncogene in genomic DNA, using droplet-based microfluidics and digital PCR.

Key words Droplet microfluidics, Digital PCR, Biomarkers, *KRAS*, Cancer

1 Introduction

A complex array of genetic alterations exists in most human cancer cells: deletions, amplifications, point mutations, and chromosomal rearrangements all play a major role in the development and progression of cancers [1, 2]. Somatic mutations within tumoral DNA can be used as highly specific biomarkers to distinguish cancer cells from their normal counterparts [3]. Mutations within tumor cells and in DNA released by tumor cells into clinical samples, such as blood, lymph, stools, or urine, can be used to detect cancers [4, 5]. Furthermore, tumor-specific genetic changes can have a profound impact on clinical decision-making and outcome [5, 6].

These DNA biomarkers are potentially useful for the diagnosis, prognosis, treatment, and follow-up of patients. In order to have the required sensitivity and specificity to detect rare tumoral DNA in stool, blood, lymph, and other patient samples, a simple, sensitive, and quantitative procedure to measure the ratio of mutant to wild-type genes is required. In order to use tumor-specific somatic mutations as biomarkers for clinical oncology, the mutations must be detected in a large excess of nonmutated DNA from normal cells.

The majority of genetic tests are based on the polymerase chain reaction (PCR), which is the most sensitive method for the detection of rare species in a complex sample. Highly sensitive methods

Valérie Taly et al. (eds.), *Microchip Diagnostics: Methods and Protocols*, Methods in Molecular Biology, vol. 1547,
DOI 10.1007/978-1-4939-6734-6_12, © Springer Science+Business Media LLC 2017

have been developed to preferentially enrich mutant sequences present at low concentrations in a background of wild-type DNA in clinical samples and selectivity of up to 10^{-9} has been described [7]. However, such assays are qualitative, which brings two disadvantages: first, an assay can yield a false-negative result (inadequate sensitivity) because the amount of starting DNA is too low to detect rare mutations. Second, an assay can yield a stochastic false-positive result (inadequate specificity) because, for example, rare random mutations are present in a sample.

Quantitative technologies, however, can overcome these problems. They have the ability to directly measure the number of DNA molecules tested per assay and therefore ensure that the amount of starting material is sufficient to detect rare mutations in the predicted frequency range. Quantification also allows random and pathogenic mutations to be distinguished by establishing a baseline mutant-to-wild-type ratio (the background mutation frequency of human cells), as well as a standard quality control monitoring which is a necessary requirement for routine diagnostic testing [8]. Most quantitative assays are analog in nature: an average signal is acquired from the mutant and wild-type DNA molecules present in the sample and the ratio between the mutant and wild-type signal is an estimate of the mutation frequency [3, 9, 10]. However, commonly used analog assays have limited sensitivity. Frequently used techniques such as pyrosequencing [11–13], or real-time PCR [14, 15] have sensitivities ranging from 1 to 10 % for the detection of mutant gDNA diluted in wild-type gDNA (see [16] for a comparative study).

More precise and sensitive quantification of mutated DNA is possible using digital techniques, which allows the analysis of many individual DNA molecules [3]. One of the most important techniques is digital PCR, which is based on the compartmentalization and amplification of single DNA molecules [17, 18]. Digital PCR allows the discrete counting of the mutant and wild-type alleles present in a sample. Its sensitivity is only limited by the number of molecules that can be analyzed and the false positive rate of the mutation detection assay. Digital assays are particularly well suited for the analysis of clinical samples like stool or plasma where tumor-derived DNA fragments represent only a small fraction of the total DNA [19, 20] and digital PCR has been used for the detection of *KRAS* mutations in various clinical samples [18, 20, 21].

The original microtitre plate-based digital PCR techniques were, however, both slow and expensive. To address these issues, several miniaturized methods allowing millions of single-molecule PCR reactions to be performed in a single assay have been developed [3]. Single-molecule PCR in a polyacrylamide film can be used to generate discrete DNA colonies (polonies) [22]. Alternatively, single-molecule PCR reactions can be performed on beads in picotiter plates [23], or in aqueous droplets in water-in-oil emulsions (BEAMing) [24]. After recovering the beads, hybridization of fluorescently labeled oligonucleotides [24] or single base extension (SBE) with fluorescently labeled dideoxynucleotides [25] is used for sequence differentiation and the beads are analyzed by flow cytometry. The resulting highly sensitive digital procedure has found many applications including the reverse-transcription PCR of RNA from extracellular vesicles [26], or detection and enumeration of rare genetic mutations [24]. For example, BEAMing followed by SBE and flow cytometry was used to quantify the level of mutated DNA circulating in the plasma of colorectal cancer patients [27]. The detection limit, determined by the error rate of the DNA polymerase used for the pre-amplification of the plasma DNA, was one mutant DNA molecule in a background of 10,000 wild-type DNA molecules.

Here, we describe a procedure allowing the highly sensitive detection of mutated DNA in a quantitative manner within complex mixtures of DNA (see also [28]). The method is based on using a droplet-based microfluidic system to perform digital PCR in millions of picolitre droplets. Genomic DNA (gDNA) is compartmentalized in droplets at a concentration of less than one genome equivalent per droplet together with two TaqMan probes, one specific for the mutant and the other for the wild-type DNA, which generate green and red fluorescent signals, respectively. After thermocycling, the ratio of mutant to wild-type genes is determined by counting the ratio of green to red droplets. We demonstrate the accurate and sensitive quantification of mutated *KRAS* oncogene in gDNA. The technique enabled the determination of mutant allelic specific imbalance (MASI) in several cancer cell lines and the precise quantification of a mutated *KRAS* gene in the presence of a 200,000-fold excess of unmutated *KRAS* genes. The sensitivity is mainly limited by the number of droplets analyzed (see [29] for a recent review on droplet-based digital PCR applications to cancer research).

Chapter 11 describes an example adaptation of this technique for a parallel multiple mutation detection where we were able to screen a gDNA sample for the presence of the six most common mutations of the *KRAS* oncogene.

2 Materials

2.1 Cell Lines

The cell lines used in this project are represented in Table 1.

2.2 Genomic DNA Preparation

1. QIAamp DNA mini Kit (Qiagen, Courtaboeuf, France).
2. Sterile nuclease-free water (Roth, Karlsruhe, Germany).
3. Nanodrop ND-1000 spectrophotometer (Thermo Scientific, Wilmington, USA) for absorbance measurements.

Table 1
Cell-lines used

Cell-lines	KRAS mutation	Culture medium	ATCC code
H358	G12C	RPMI-1640	CRL-5807
LS123	G12S	EMEM	CCL-225
SW620	G12V	Leibovitz's L-15	CCL-227
H1573	G12A	DMEM	CRL-5877
LoVo	G13D	F-12K	CCL-229
SW48	Wild-type	DMEM	CCL-231

2.3 Materials for Droplet Microfluidics

1. PHD 2000 (Harvard Apparatus, Les Ulis, France) syringe pumps.
2. A home-made Labview (National Instruments) interface.
3. 1 mL polyethylene syringes (Omnifix- FR, BRAUN, Bad Arolsen, Germany).
4. 23G syringe needles.
5. 0.56 mm × 1.07 mm polytetrafluoroethylene (PTFE) tubings.
6. HFE-7500 oil (3M) with 2 % (w/w) EA-surfactant (Raindance, Lexington MA, USA).

2.4 Microfluidics Device Preparation and Operation

1. AutoCAD, (Autodesk).
2. Photolithography Mask (Selba Corp, Switzerland).
3. PDMS and curing agent (Sylgard® 184 silicone elastomer kit, Dow Corning).
4. SU8 photoresist and developer (MicroChem Corp.).
5. Silicon wafers (Siltronix).
6. MJB3 contact mask aliner; (SUSS MicroTech).
7. 0.75 mm Uni Core biopsy puncher (Ted Pella Inc., USA).
8. Oxygen plasma cleaner (Diener Electronic, Ebhausen, Germany).
9. 2 % perfluorooctyldimethylsilane (Aldrich, Sigma-Aldrich ChimieS.a.r.l., Lyon, France).
10. Fluorinated oil (HFE-7500, 3M) containing 2 % (w/w) EA surfactant (a PEG-PFPE amphiphilic block copolymer [30].

2.5 TaqMan Probes and Primers

1. The TaqMan® MGB probes (Applied Biosystems, Life Technologies corp., Ville-bon sur Yvette, France) and the primers allowing amplification of the *KRAS* target sequences (Eurogentec Angers, France) were previously developed and

validated for clinical sample analysis [31]. The probes specific for mutated sequences are conjugated to the fluorophore 6-FAM (λex 494 nm/λem 522 nm), whereas the probes complementary to wild-type sequence are conjugated to the fluorophore NED (λex 546 nm/λem 575 nm).

2.6 PCR Reagents

1. Detergent-free buffer (Finnzymes, Espoo, Finland).
2. Dextran Texas Red (DTR) dye (Life Technologies).
3. dNTPs mix (MP Biomedicals).
4. DyNAzyme II polymerase (5 U/μL, Finnzymes, Espoo, Finland).

2.7 The Collection-Reinjection Device

1. 0.5 mL PCR tube (Thermo Fisher Scientific, Illkirch, France).
2. PDMS plug (created with an 8 mm diameter biopsy punch from a 6 mm thick PDMS disk pre-poured in a Petri dish).

2.8 On-Chip Droplet Analysis and Fluorescence Detection

2.8.1 Optical Setup

1. Axiovert 200 inverted microscope (Carl Zeiss SAS).
2. Vibration-dampening platform (Vibraplane, Kinetic Systems).
3. Solid-state lasers (25 mW) of 488 nm (Green) and 532 nm (Blue) wavelength (Newport-Spectraphysics) Heatsinks (Newport-Spectraphysics).
4. Dichroicbeam splitter (LM01-503-25; Semrock Inc.).
5. Periscope assemblies (Thorlabs GmbH).
6. Microscope objective (Carl Zeiss SAS).
7. Bandpass filters: 562/40 BrightLine (Semrock Inc.), F1 (FF01-617/73-25; Semrock), F2 (FF01- 510/20-25; Semrock).
8. 488 nm/532 nm/638 nm-wavelength transmitting dichroic beam splitter (Di-T488/532/638-25x36x5.0; Semrock).
9. Notch filters (NF01-488-25 and NF01 532U-25; Semrock).
10. A single-edge dichroic beam splitter (FF562-Di02- 25x36; Semrock).
11. PMTs: Orange (NED) (H5784-20; Hamamatsu Photonics KK) and Green (FAM) PMT (H5784-20; Hamamatsu Photonics KK).
12. PCI-7831R Multifunction Intelligent data acquisition (DAQ) card (National Instruments Corporation) executing a program written by in LabView 8.2 (FPGA module, National Instruments Corporation).

3 Methods

3.1 Cell Culture

All cell culture medium were purchased from ATCC and supplemented with 10 % Fetal Bovine Serum (GIBCO-BRL) and 1 %

penicillin/streptomicyn (GIBCE-BRL). Cells were incubated at 37 °C under a 5 % CO2 atmosphere saturated with water. For passaging cells were trypsinized (0.25 % trypsin in PBS, GIBCO-BRL) and a fraction 1/10 of the resulting suspension was seeded into new culture flasks. Subsequently fresh medium was added.

3.2 Genomic DNA Preparation (See Note 1)

1. Extract Genomic DNA (gDNA) from cultured cells using a QIAamp DNA mini Kit (Qiagen, Courtaboeuf, France), following manufacturer's protocol.
2. Elute gDNA was in sterile nuclease-free water (Roth, Karlsruhe, Germany).
3. Determine purified DNA concentration by measuring the absorbance at 260 nm using a Nanodrop ND-1000 spectrophotometer (Thermo Scientific, Wilmington, USA).

3.3 Microfluidics Device Preparation

1. Microfluidic structures were designed using a computer-assisted drawing program (AutoCAD, Autodesk) and were sent to Selba Tech to be printed as mirrored negative photomasks at a resolution of 25,400 dpi.
2. Once the photolithography mask is obtained pattern the channels in poly(dimethylsiloxane) (PDMS) using conventional soft lithography techniques [32].
3. According to the device depth spin coat SU8 photoresist (MicroChem Corp) onto silicon wafers (Siltronix) and pattern by UV exposure (MJB3 contact mask aliner; SUSS MicroTech) through the photolithography mask.
4. After the exposure spin coat the silicon wafer with the subsequent SU8 developer (MicroChem Corp).
5. Prepare the PDMS and curing agent (Dow Corning Corp) as a mixture of 10:1 (w/w).
6. Degas the mixture under vacuum and pour onto the silicon wafer and place the wafer in a 65 °C incubator for 2 h for curing.
7. Peel the cured PDMS from the master and produce input/output ports by punching with a 0.75 mm Uni Core biopsy punch.
8. Bond the structured side of the PDMS onto a 1 mm thick glass slide using an oxygen plasma (Diener Electronic, Ebhausen, Germany).
9. To decrease the wettability of the aqueous phase on the channel walls inject 2 % perfluorooctyldimethylsilane (Aldrich, Sigma-Aldrich Chimie, Lyon, France) in HFE-7100 oil (3M, St Paul, USA). Flush the excess fluorosilane out with Nitrogen.

3.4 Microfluidics Device Operation (See Note 2)

1. The aqueous solution containing the PCR mix and gDNA was placed in a 1 mL polyethylene syringe (Omnix-F®, BRAUN, Bad Arolsen, Germany) with a 23G syringe needle and injected into the dropmaker device (Fig. 1) using 0.56 mm × 1.07 mm polytetrafluoroethylene (PTFE) tubing.
2. Fluorinated oil (HFE-7500, 3M) containing 2 % (w/w) EA surfactant (a PEG-PFPE amphiphilic block copolymer [27] was placed in a 1 mL syringe.
3. The aqueous phase was injected at 100 μL/h and the oil phase at 200 μL/h using syringe pumps (PHD 2000, Harvard Apparatus, Les Ulis, France) and 10 pL droplets were generated on-chip by hydrodynamic flow-focusing [33] (a production frequency of 3 kHz).
4. The produced emulsion was collected in a "collection-reinjection device" for thermocycling.

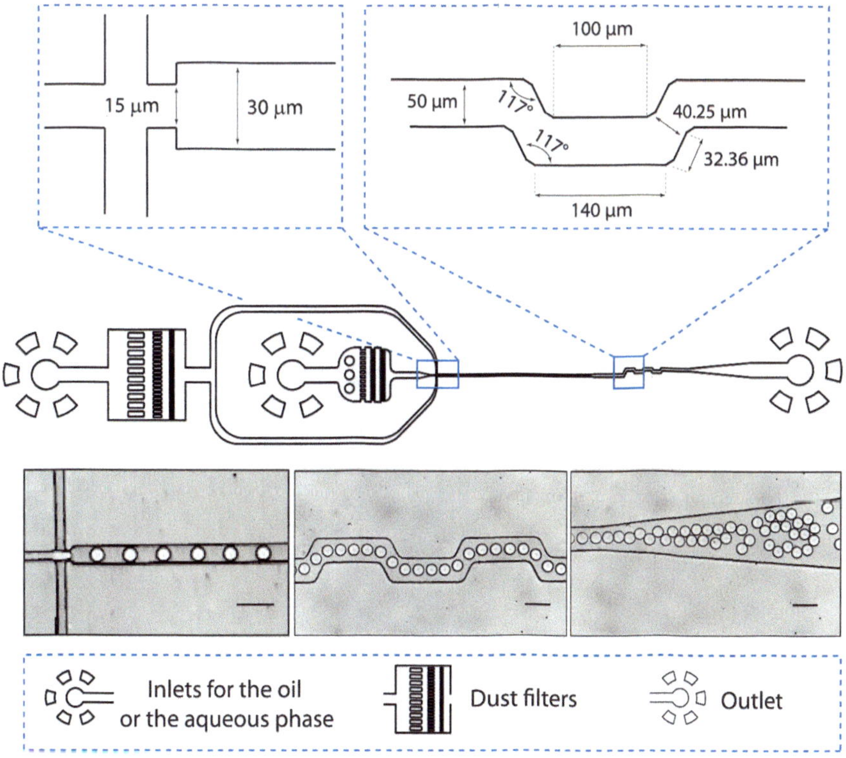

Fig. 1 Microfluidic device for the droplet production. The device contains one inlet for the oil, one inlet for the aqueous phase, a flow focusing connection (*top left panel*), and a mixing module (*top right panel*). The *bottom panel* represents a light micrograph showing the drop production, the mixing, and droplet collection (Scale bar 60 μm) (Figure modified from [28], permission from Lab on Chip)

3.5 The Collection-Reinjection Device

1. Cut a PDMS plug using a 8 mm biopsy puncher.
2. Create two holes on the plug with a 1 mm biopsy punch (Fig. 2).
3. Insert the PDMS plug inside the 0.5 mL PCR tube. The position of the plug inside the tube is determined so that all the emulsion plunges inside the thermocycler well.
4. Fill the device entirely with oil taking care not to leave any air bubbles inside.
5. Use 0.56 mm × 1.07 mm PTFE tubings to ensure proper emulsion collection efficient reinjection.

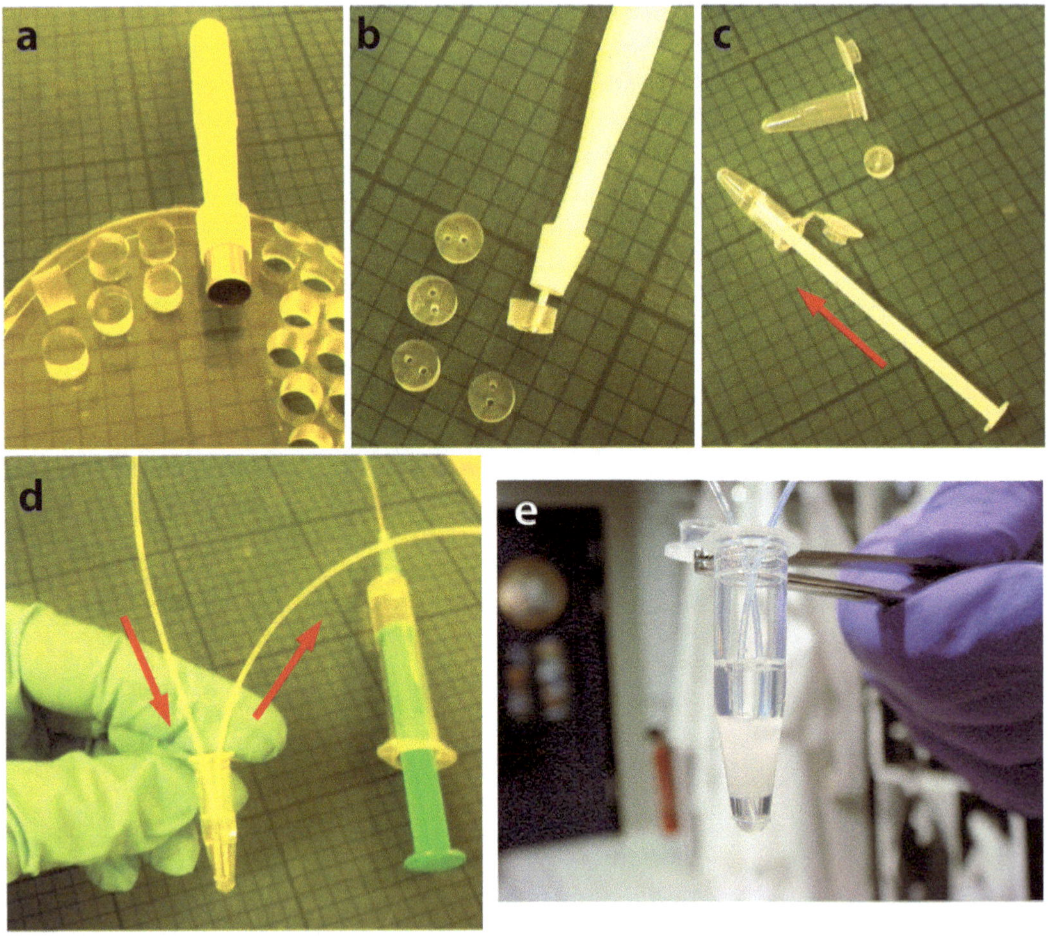

Fig. 2 Collection-reinjection device: A common 8 mm biopsy needle is used to punch plugs from previously prepared PDMS, cured 10:1 (**a**) and a 1 mm biopsy needle is used to punch two holes in the plug for the in/out tubings (**b**). The plug is then inserted into a 0.5 mL PCR tube by pushing with a syringe plunger (**c**). A tubing, connected to a syringe filled with oil, is inserted from one of the holes on the PDMS plug to the bottom of the tube and a second tubing is inserted from the other hole to the end of the PDMS plug (**d**). The vial is filled with oil from the bottom to the top. The second tubing helps the air bubble to come out. It is important at this step to not leave any air inside the tube since the expansion of the air during thermocycle will press the droplets against each other and cause coalescence

3.6 Thermocycling of Emulsions

1. After the removal of the tubings, the device containing the emulsion was placed in a PCR Machine (Mastercycler®, Eppendorf, Le Pecq, France) and the emulsion was thermocycled, starting with an initial denaturation step of 2 min at 94 °C, followed by 39 cycles of: 94 °C, 15 s; 60 °C, 1 min.

3.7 On-Chip Fluorescence Detection

1. The optical setup used for the experiments is already described in reference (Fig. 3).
2. An Axiovert 200 inverted microscope (Carl Zeiss SAS) mounted on a vibration-dampening platform (Vibraplane, Kinetic Systems).

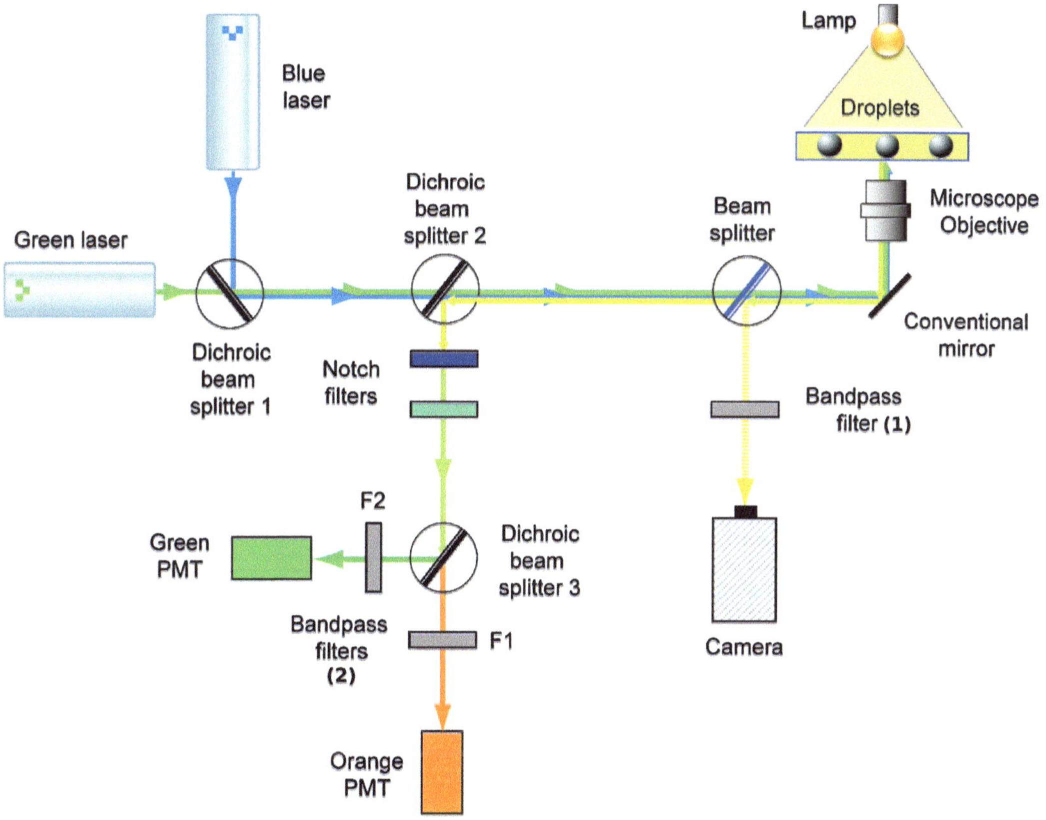

Fig. 3 The optical setup. An Axiovert 200 inverted microscope (Carl Zeiss SAS) mounted on a vibration-dampening platform (Vibraplane, Kinetic Systems) was used. Two solid-state lasers (25 mW) of 488 nm (*Green*) and 532 nm (*Blue*) wavelengths (Newport-Spectraphysics) were fixed on the platform via heatsinks (Newport-Spectraphysics). The beams of the 488 nm and 532 nm lasers were shaped into the 10 × 150 μm wide line and a 20 μm wide spot, respectively. The two laser beams were combined detection using a dichroic beam splitter (LM01-503-25; Semrock Inc.) and guided to the side camera port of the microscope via a series of periscope assemblies (Thorlabs GmbH). The laser light was reflected up into an LD Plan Neofluar 40×/0.6 microscope objective (Carl Zeiss SAS) and focused in the microfluidic channel. The bandpass filter (1) is a 562/40 BrightLine bandpass filter (Semrock Inc.) and is positioned in front of the camera to protect the camera sensor from the laser light (Figure modified from [34], permission from Lab on Chip)

3. Two solid-state lasers (25 mW) of 488 nm (Green) and 532 nm (Blue) wavelength (Newport-Spectraphysics) were fixed on the platform via heatsinks (Newport-Spectraphysics).
4. The beams of the 488 nm and 532 nm lasers were shaped into the 10 × 150 μm wide line and a 20 μm wide spot, respectively.
5. The two laser beams were combined detection using a dichroic beam splitter (LM01-503-25; Semrock Inc.) and guided to the side camera port of the microscope via a series of periscope assemblies (Thorlabs GmbH).
6. The laser light was reflected up into an LD Plan Neofluar 40×/0.6 microscope objective (Carl Zeiss SAS) and focused in the microfluidic channel.
7. A bandpass filter 562/40 BrightLine (Semrock Inc.) is positioned in front of the camera to protect the camera sensor from the laser light.
8. A Phantom v4.2 high-speed digital camera (Vision Research) was mounted on the top camera port of the microscope to capture digital images during droplet production, fusion, and reinjection. A 562/40 BrightLine bandpass filter (Semrock Inc.) positioned in front of the camera protected the cameras sensor from reflected laser light.
9. Light emitted from fluorescent droplets was captured by the objective and channeled back along the path of the laser into the system of periscope assemblies. The emitted light was separated from the laser beam by a 488 nm/532 nm/638 nm-wavelength transmitting dichroic beam splitter (Di-T488/532/638-25x36x5.0; Semrock), filtered through the two notch filters (NF01-488-25 and NF01-532U-25; Semrock) and split between two photomultiplier tubes (PMTs) by a single-edge dichroic beam splitter (FF562-Di02- 25x36; Semrock).
10. Transmitted light passed through a bandpass filter (FF01-617/73-25; Semrock) to the Orange (NED) PMT (H5784-20; Hamamatsu Photonics KK). Reflected light passed through another bandpass filter F2 (FF01- 510/20-25; Semrock) to the Green (6-FAM) PMT (H5784-20; Hamamatsu Photonics KK).
11. The signal output from the PMT was analyzed using a PCI-7831R Multifunction Intelligent data acquisition (DAQ) card (National Instruments Corporation) executing a program written by in LabView 8.2 (FPGA module, National Instruments Corporation), which allowed the identification of droplets by peaks in fluorescence, as well as the width of each droplet. The data acquisition rate for the system was 100 kHz.

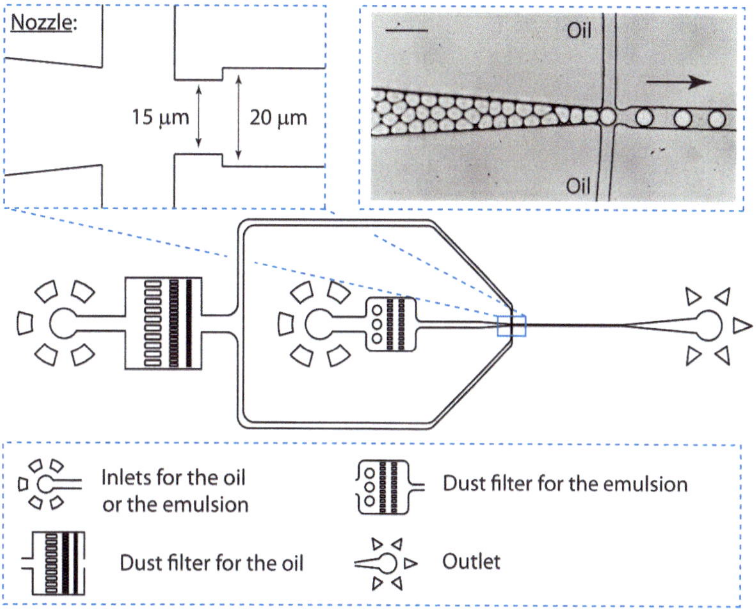

Fig. 4 Reinjection device. The device contains one inlet for the oil and one inlet for the emulsion. *Left Panel*: dimensions of the nozzle. *Right Panel*: a light micrograph showing the spacing of the reinjected emulsion by oil (Scale bar 60 μm, Figure modified from [28], permission from Lab on Chip)

3.8 Droplet Fluorescence Analysis

1. Droplets were reinjected in a microfluidic reinjection device (Fig. 4) and spaced with HFE-7500 oil in a 30 μm wide 25 μm deep channel. Droplet fluorescence was measured in the middle of the channel using the optical setup described in Fig. 3.
2. For each droplet passing through the channel the fluorescence (as a green or orange peak) and the droplet width (the time it takes for a droplet to pass through the laser beam) was analyzed.
3. The fluorescent signal was then converted from volts (V) to Relative Fluorescence Units (RFU) according to the equation: $RFU = U/G\ 7.2$, where U is the height of the fluorescence peak measured by PMT voltage output and G the PMT gain input (V) while the exponent 7.2 is given by the PMT manufacturer (Hamamatsu Photonics). The diameter D of the droplets was determined using the droplet production frequency either with fluorescence analysis or directly using the high-speed camera according to the equation $W = F/f$ where W is the droplet width in picoliters, F is the flow rate in μL/sec, and f is the droplet production frequency calculated as the number of fluorescence peaks per second or counted using the high-speed camera.

3.9 Confocal Microscopy Imaging

1. The emulsion was spread between two 0.17 mm thick microscope coverslips and images were acquired with LSM510 laser-scanning confocal microscope (Zeiss, Jena, Germany) equipped with C-Apochromat 20× (n.a. 0.8) water immersion objectives in multitrack mode.

2. The 6- FAM fluorophore was excited at 488 nm using an argon laser, and the emission spectra were recorded in the range 505–550 nm.
3. The NED fluorophore was excited using a helium-neon 561 nm laser and the emission spectra were recorded beyond 585 nm.
4. Images were processed with Zeiss LSM Image Browser, version 4.0SP1, software and with ImageJ 1.42q (National Institutes of Health, USA) to obtain composites.

3.10 Mutation Detection Sensitivity

1. To estimate the sensitivity of the procedure, Genomic DNA from the heterozygous cell-line H1573, which contains a G to C transversion in the second base of *KRAS* codon 12, was serially diluted into gDNA containing the wild-type *KRAS* gene from the cell-line SW48.
2. Both gDNA (H1573 and SW48) were fragmented by nebulization to obtain 1–5 kb fragments.
3. The dilutions were performed from 1/10 to 1/100,000, leading to ratios of mutant/wild-type *KRAS* alleles from ~1/20 to ~1/200,000 (the initial fraction of the mutant allele in the H1573 cell-line is 49 %) (Fig. 5).
4. A 200 μL PCR reaction was set up containing: DyNAzyme detergent-free buffer (Finnzymes, Espoo, Finland); 200 μM dNTPs (MP Biomedicals); 0.3 μM forward primer; 0.3 μM reverse primer (Eurogentec, Angers, France); 200 nM of TaqMan® probe specific of mutant sequence; 750 nM TaqMan® probe specific of wild-type sequence; 5 units of DyNAzyme II polymerase (5 U/μL, Finnzymes, Espoo, Finland).
5. The DNA mixtures were diluted to 30 ng/μL, and compartmentalized in 9 pL droplets, together with PCR reagents, primers, and TaqMan® probes specific for both mutant and wild-type DNA.
6. After the thermocycle, droplets were analyzed by on-chip droplet fluorescence detection (Fig. 6).

3.11 Mutant-Allele Specific Imbalance (MASI) Analysis

1. Genomic DNA extracted from H1573, H358, LoVo, and LS123 heterozygous cell-lines was used for the assessment of mutant to wild-type *KRAS* alleles.
2. A duplex assay (as described above) was used with 6 μg of gDNA from each cell-line with droplet sizes of 9 pL for H358, LoVo, SW620 and SW48 cell-lines, and 4 pL for LS123 and H1573 cell-lines.
3. After the thermocycle, the droplets were analyzed with confocal microscopy imaging. The ratio of mutant to wild-type *KRAS* alleles (mA%) was determined by counting the number of green fluorescent droplets over the number of red fluorescent droplets.

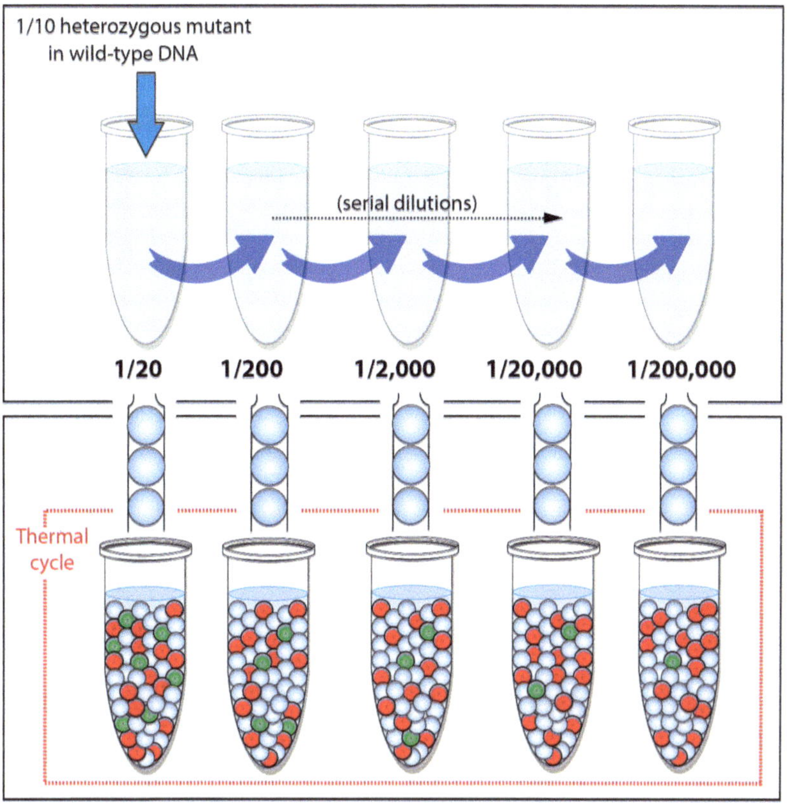

Fig. 5 The protocol for the sensitivity test. A ten-fold serial dilution of genomic DNA extracted from the *KRAS* mutant H1573 cell-line was realized inside genomic DNA from *KRAS* wild-type SW48 cell-line from 1/10 (mutant/wild-type) to 1/100,000. Since the H1573 cell-line is a heterozygous mutant this corresponded to a serial dilution of 1/10 to 1/200,000. Each step was submitted to the droplet-based microfluidic PCR procedure and analyzed separately. The dilutions from 1/20 to 1/20,000 were performed as duplicates and the 1/200,000 dilution was performed in triplicate

3.12 Results

3.12.1 Emulsion Stability After Thermocycling

The emulsion stability during thermocycle in the collection-reinjection tube was assessed by comparing the droplet widths before and after thermocycle. A uniform distribution of the droplet width after thermocycle was considered a sign of a noncoalesced emulsion. The results of this test are represented in Table 2, showing that after thermocycling only less than 2 % of the droplets coalesced. The same experiment was repeated without adding Fluorescein into the PCR mix and the droplet integrity was evaluated visually by reinjecting the emulsion in a microfluidic reservoir (Fig. 7).

3.12.2 MASI Analysis by Confocal Microscopy Imaging

A Poisson distribution of gDNA was expected in the droplets and, assuming that the mass of a (haploid) genome equivalent is 3.3 pg, in an emulsion containing 100 μL of aqueous phase each droplet should contain, on average, 0.08 genome equivalents (for the 9 pL droplets) or 0.04 genome equivalents (for the 4 pL droplets).

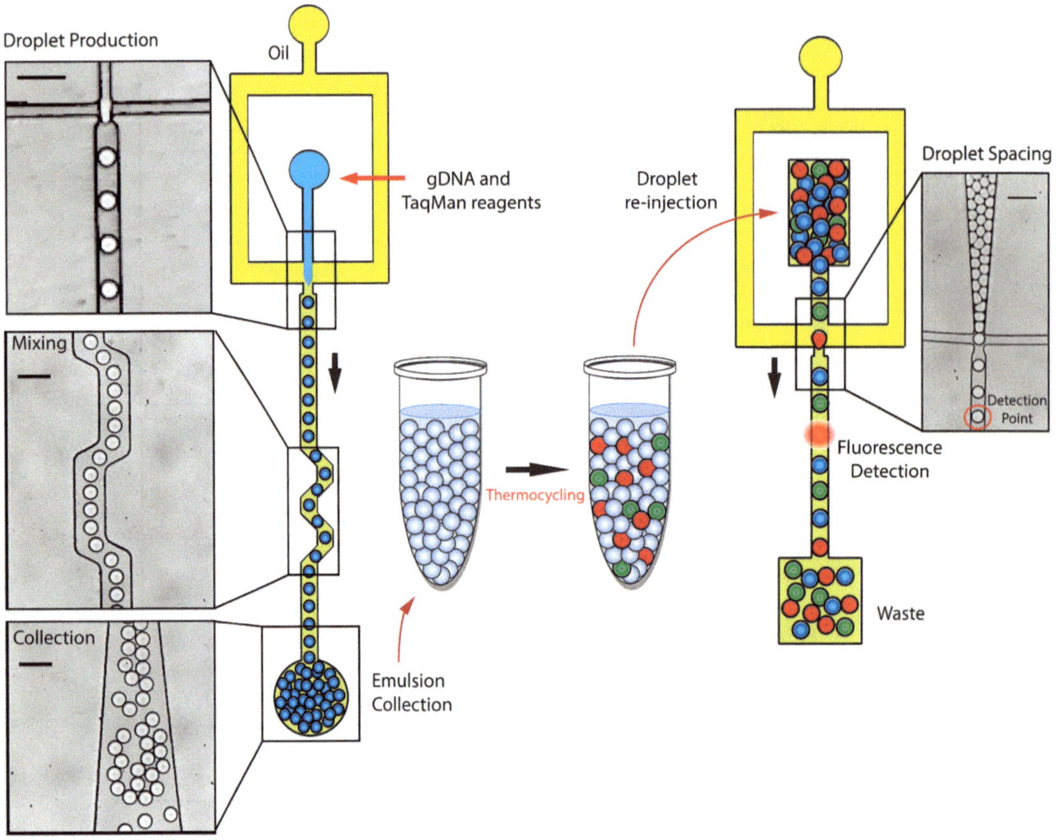

Fig. 6 Overview of the system. An aqueous phase containing the gDNA, PCR reagents, and TaqMan® probes specific for the wild-type and mutant genes is emulsified within a microfluidic device. Light micrographs of drop production, mixing, and collection are shown (scale bars 60 μm). The concentration of gDNA is such that there is, on average, less than one (haploid) genome equivalent per droplet. The emulsion is collected and thermocycled. During DNA amplification, the TaqMan® probes are cleaved and the corresponding fluorophores are released. TaqMan® probes specific for the mutant sequence carry a 6-FAM fluorophore (λex 494 nm/λem 522 nm) and TaqMan® probes specific for the wild-type sequence carry a NED fluorophore (λex 546 nm/λem 575 nm). Thus, the amplification of mutant DNA gives green-fluorescent droplets while the amplification of wild-type DNA gives *red-fluorescent droplets*. The emulsion is then reinjected onto a microfluidic chip, the droplets are spaced by oil, and the fluorescent signal of each droplet is analyzed. A light micrograph of reinjection and spacing is shown (scale bar 60 μm, Figure modified from [28], permission from Lab on Chip)

Table 2
Coalescence analysis for the collection-reinjection tube

	Number of droplets	Coalesced droplets (%)	SD main peak
Production[bt]	241,033	0.05 % (SD = 0.0007)	0.006 (SD = 0.0018)
Reinjection[at]	184,400	1.92 % (SD = 0.0055)	0.082 (SD = 0.0288)

A PCR mix with all the components necessary for a PCR reaction (except for target DNA) with 1 μM Fluorescein was emulsified in 10 pL droplets and thermocycled. Droplet width before thermocycle (bt) and after the thermocycle (at) was measured and compared. All the experiments were performed in triplicate and the results are represented as the average of the data collected in the three experiments (standard deviation (SD) given in brackets)

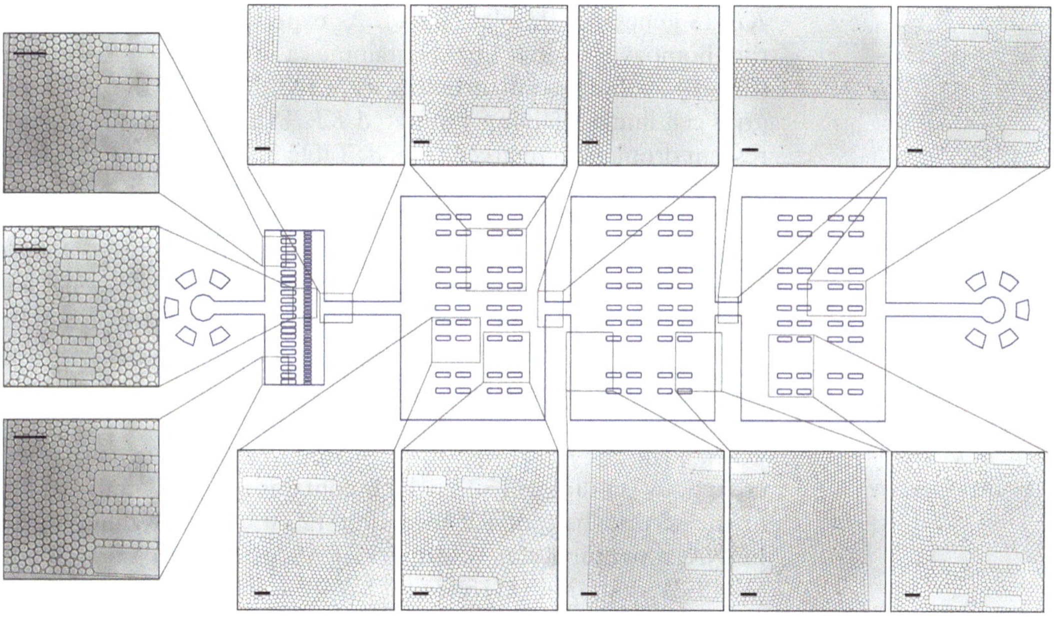

Fig. 7 Microfluidic reservoir for the visual assessment of the emulsion stability

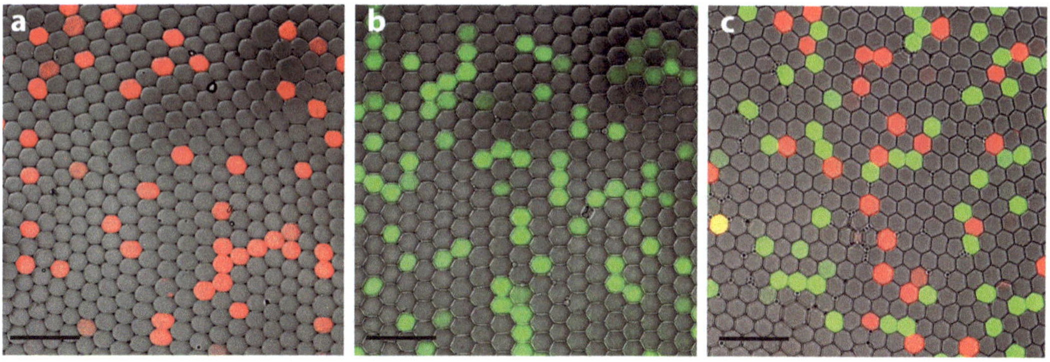

Fig. 8 Fluorescence confocal microscopy analysis of thermocycled droplets. gDNA extracted from homozygous cell lines bearing wild-type *KRAS* alleles (SW48) (**a**), mutant *KRAS* alleles (SW620) (**b**), and a heterozygous cell-line bearing both mutant and wild-type *KRAS* alleles (H358) (**c**) were analyzed by fluorescence confocal microscopy. TaqMan® probes specific for the wild-type sequence carried a NED fluorophore (λex 546 nm/λem 575 nm) and TaqMan® probes specific for the mutant sequences carried a 6-FAM fluorophore (λex 494 nm/λem 522 nm). *Red-fluorescent droplets* contain wild-type DNA, *green-fluorescent droplets* contain mutant DNA, *yellow-fluorescent droplets* contain both mutant and wild-type DNA, and nonfluorescent droplets do not contain target DNA (scale bar 100 μm) (Figure modified from [28], permission from Lab on Chip)

For H358, LoVo, SW620, and SW48 cell-lines the ratio of fluorescent to nonfluorescent droplets was in the range of 0.06–0.1, consistent with the expected number of *KRAS* genes per droplet (0.08) (Fig. 8). For LS123 and H1573 cell-lines this ratio was in the range of 0.03–0.04, consistent with the expected number of

KRAS genes per droplet (0.04). As expected, with gDNA from the homozygous cell-line containing a wild-type *KRAS* gene (SW48) all fluorescent droplets were red and with the homozygous cell-line containing a mutated *KRAS* gene (SW620) all fluorescent droplets were green (Fig. 8, Table 3). For the four different heterozygous cell-lines containing a mutated and a wild-type *KRAS* gene (H1573, H358, LoVo and LS123), the fluorescent droplets were either red (due to amplification of a wild-type *KRAS* gene), green (due to amplification of a mutant *KRAS* gene), or yellow (due to amplification of a mutant and a wild-type *KRAS* gene). The MASI determined was in good agreement with the previously published data [35] (Table 3).

3.12.3 Mutation Detection Sensitivity

To estimate the sensitivity of the procedure, gDNA from the heterozygous cell-line H1573, which contains a G to C transversion in the second base of KRAS codon 12, was serially diluted into gDNA containing the wild-type KRAS gene from the cell-line SW48. Both gDNA (H1573 and SW48) were fragmented by nebulization to obtain 1–5 kb fragments. The dilutions were performed from 1/10 to 1/100,000 leading to ratios of mutant/wild-type KRAS alleles from ~1/20 to ~1/200,000 (the initial fraction of the mutant allele in the H1573 cell-line is 49 %) (Fig. 9). This allowed to vary the ratio of mutant *KRAS* to wild-type *KRAS* over five decades while the ratio of droplets containing wild-type *KRAS* genes over the total number of droplets was kept constant (0.08). Four types of droplets were then generated:

- red-fluorescent droplets: containing wild-type *KRAS* genes
- green-fluorescent droplets: containing mutant *KRAS* genes
- nonfluorescent droplets: that do not contain *KRAS* genes
- yellow fluorescent droplets: containing both wild-type and mutant genes as a result of the combination of green and red fluorescence.

As mentioned earlier, it has been shown that the encapsulation of single gene in droplets is described by a Poisson distribution [28, 36]. As such for the encapsulation of km mutant genes and k_{wt} wild-type genes in a single droplet are given by the product of the Poisson distribution of the two independent random variables k_m and k_{wt} (Eq.1):

$$P(k_m, k_{wt}) = \frac{e^{-\lambda m}\lambda_m^{km}}{k_m!} x \frac{e^{-\lambda_{wt}}\lambda_{wt}^{k_{wt}}}{k_{wt}!} \tag{1}$$

The statistics of encapsulation are presented in Fig. 10 for clarity for three different cases (λ_{wt} = 0.8 (a); 0.08 (b); and 0.008 (c)). For high λ_{wt} (Fig. 10a) the number of yellow droplets has to be evaluated to count all the mutant and wild-type genes.

Table 3
Mutant allelic specific imbalance (MASI) analysis (Figure modified from [28], permission from Lab on Chip)

		Number of droplets	Number of droplets containing DNA				Reported mA(%)			
Cell line	Mutation	analyzed	Green droplets	Red droplets	Yellow droplets	Experimental mA %	DNA seq.	RFLP	Subcloning	cDNA seq.
SW48	–	9120	0	1080	0	0	NA	NA	NA	NA
SW620	G12V	4938	621	0	0	100	NA	NA	NA	NA
H1573	G12A	5612	119	116	2	49.3	NA	NA	NA	NA
H358	G12C	4677	239	145	0	62.2	68.7	85.0	82.4	71.8
LoVo	G13D	28905	2317	1134	34	67.1	66.7	NA	68.4	NA
LS123	G12S	23192	336	330	9	50.4	60.3	51.3	55.0	NA

Variations in the ratio of mutant to wild-type KRAS alleles (mA%) were determined in four heterozygous cell-lines (H1573, H358, LoVo, and LS123) and compared with literature data obtained by DNA sequencing (DNA seq.), restriction fragment length polymorphism (RFLP), subcloning or cDNA sequencing (cDNA seq.).[30] No literature data were available (NA) for the H1573 cell-line. The cell-lines SW48 and SW620, ***which are homozygous for wild-type and mutant KRAS, respectively, were used as controls.***

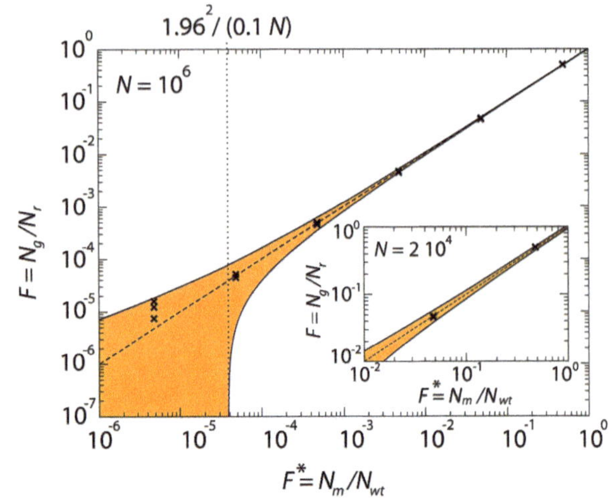

Fig. 9 Determination of the sensitivity. Experimental measurement of the fraction of green-fluorescent droplets containing mutant *KRAS* (*Ng*) over the number of *red-fluorescent droplets* containing wild-type *KRAS* (*Nr*) as a function of the ratio of mutant (N_m) to wild-type (N_{wt}) genes. In an ideal assay, *Ng*/*Nr* is equal to N_m/N_{wt} (*dashed line*). The 95 % confidence interval (*orange area*) for the analysis of $N = 10^6$ droplets when $N_{wt}/N = 0.1$ is shown. All the experimental points (x) obtained in duplicate (except for the 1/200 000 which was in triplicate) for the different dilutions fall into this 95 % confidence interval. The dotted line corresponds to the point where a measured $N_m = 0$ falls in the 95 % confidence interval. Inset: higher ratios of mutant to wild-type genes, statistically relevant data are obtained by the analysis of a smaller subset of droplets. The experimental points (x) are plotted as a zoom and compared to the 95 % confidence interval (*orange area*) determined from the analysis of $N = 2.10^4$ droplets analyzed experimentally (Figure modified from [28], permission from Lab on Chip)

When a very low λ_{wt} is used ($\lambda_{wt} = 0.008$), the yellow droplets appear insignificant in the statistics (Fig. 10c). However in this case, the experiment requires the analysis of a majority of empty droplets, which reduces the throughput of the test. Finally for intermediate values ($\lambda_{wt} = 0.08$, Fig. 10b), the yellow droplets represent less than 10 % of the mutant droplet. The mathematical modelization for the determination of the 95 % confidence was described previously [28, 36].

4 Notes

1. Measures to Minimize Contamination
   a. To minimize contamination by exogenous DNA, all DNA manipulations prior to amplification should be performed in a UV-treated environment (Biocap DNA/RNA hood, Fisher Bioblock Scientific, Illkirch, France).

(a)

λ_{WT}= 0.8000 / λ_M= 0.0800		Wild-type 0	1	2	3	4
	0	**4.1E-01**	**3.3E-01**	1.3E-01	3.5E-02	7.1E-03
	1	**3.3E-02**	2.7E-02	1.1E-02	2.8E-03	5.7E-04
Mutant	2	1.3E-03	1.1E-03	4.2E-04	1.1E-04	2.3E-05
	3	3.5E-05	2.8E-05	1.1E-05	3.0E-06	6.0E-07
	4	7.1E-07	5.7E-07	2.3E-07	6.0E-08	1.2E-08

Total Empty:	4.1E-01
Total Red:	5.1E-01
Total Green:	3.5E-02
Total Yellow:	4.2E-02
Total :	0.99858866
yellow/green:	1.22240000

(b)

λ_{WT}= 0.0800 / λ_M= 0.0080		Wild-type 0	1	2	3	4
	0	**9.2E-01**	**7.3E-02**	2.9E-03	7.8E-05	1.6E-06
	1	**7.3E-03**	5.9E-04	2.3E-05	6.3E-07	1.3E-08
Mutant	2	2.9E-05	2.3E-06	9.4E-08	2.5E-09	5.0E-11
	3	7.8E-08	6.3E-09	2.5E-10	6.7E-12	1.3E-13
	4	1.6E-10	1.3E-11	5.0E-13	1.3E-14	2.7E-16

Total Empty:	9.2E-01
Total Red:	7.6E-02
Total Green:	7.4E-03
Total Yellow:	6.1E-04
Total :	0.99999997
yellow/green:	0.08328704

(c)

λ_{WT}= 0.0080 / λ_M= 0.0008		Wild-type 0	1	2	3	4
	0	**9.9E-01**	**7.9E-03**	3.2E-05	8.5E-08	1.7E-10
	1	**7.9E-04**	6.3E-06	2.5E-08	6.8E-11	1.4E-13
Mutant	2	3.2E-07	2.5E-09	1.0E-11	2.7E-14	5.4E-17
	3	8.5E-11	6.8E-13	2.7E-15	7.2E-18	1.4E-20
	4	1.7E-14	1.4E-16	5.4E-19	1.4E-21	2.9E-24

Total Empty:	9.9E-01
Total Red:	8.0E-03
Total Green:	7.9E-04
Total Yellow:	6.4E-06
Total :	1.00000000
yellow/green:	0.00803209

Fig. 10 Probabilities of occurrence of *red, green, empty,* and *yellow droplets* for different values of λ_{wt} and a constant ratio λ_m/λ_{wt}. The background color shows the resulting fluorescence of the droplets. Droplets containing no gene appear *black*, wild-type gene *red* and mutant gene *green*. When one or more mutant genes are co-encapsulated with one or more wild-type genes the droplet appears *yellow*. See Eq. (1) for the details of the calculations. From (**a**) to (**c**) the occupancy is decreasing resulting in a decrease of the occurrence of *yellow* droplets (Figure modified from [28], permission from Lab on Chip)

b. The TaqMan® probes should be aliquoted upon arrival and stored at 20 °C. A new aliquot should be used for each set of experiments.

c. Dilutions of PCR reagents and dyes should be performed in sterile DNase-, RNase-free, DEPC-treated water (Roth, Karlsruhe, Germany).

d. A new microfluidic device should be used for each set of experiments to limit carry-over effects.
e. Carry out all the experiments using certified DNA-free, sterilized, filtered pipette tips.
f. In addition, we recommend using disposable tweezers (Roth, Karlsruhe, Germany) for microfluidic experiments. All disposables (tweezers, PCR tubes, collection-reinjection devices, tubing, etc.) should be UV-irradiated before use.

2. Measures to Minimize Coalescence
   a. Make sure there are no air bubbles remaining inside the collection-reinjection device. During the thermocycle the air bubbles will cause the emulsion to coalesce.
   b. At the final step of the gDNA extraction, the DNA should be eluted in sterile DNase-, RNase-free, DEPC-treated water (Roth, Karlsruhe, Germany) and not the conventional Tris–EDTA Buffer. Being viscous, EDTA may cause a jetting at the droplet production and hinder the emulsion stability.
   c. Detergents in the aqueous phase have a negative effect on droplet stability. Therefore, a detergent-free PCR buffer should be used.
   d. Adding 1 mg/mL Bovine Serum Albumine (BSA) may help in limiting droplet coalescence.
   e. The solution in which most polymerases are provided has a negative effect on emulsion stability. Hence, a highly concentrated enzyme solution (5 U/μL) should be preferred.

Acknowledgments

This work was supported by Région Alsace, the Ministère de l'Enseignement Supérieur et de la Recherche, the Université de Strasbourg, the Université Paris Descartes, the Centre National de la Recherche Scientifique (CNRS), the Institut National de la Santé et de la Recherche Médicale (INSERM), the Institut National du Cancer (INCa, no. 2009-1-RT-03-US-1) and the Association pour la recherche sur le Cancer (ARC, no. SL220100601375).

References

1. Vogelstein B, Kinzler KW (2004) Cancer genes and the pathways they control. Nat Med 10:789–799
2. Stratton MR, Campbell PJ et al (2009) The cancer genome. Nature 458(7239):719–724
3. Diehl F, Diaz LA (2007) Digital quantification of mutant DNA in cancer patients. Curr Opin Oncol 19(1):36–42
4. Sawyers CL (2008) Cancer biomarker problem. Nature 452:548–552
5. Lecomte T, Berger A, Zinzindohou F et al (2002) Detection of free circulating tumor-associated DNA in plasma of colorectal cancer patients and its association with prognosis. Int J Cancer 100(5):542–548
6. Li J, Wang L, Mamon H, Kulke MH et al (2008) Replacing PCR with COLD-PCR enriches variant DNA sequences and redefines the sensitivity of genetic testing. Nat Med 14(5):579–584
7. Milbury CA, Li J, Makrigiorgos MG (2009) PCR-based methods for the enrichment of minority alleles and mutations. Clin Chem 55(4):632–640
8. Caen O, Nizard P et al (2015) Digital PCR compartmentalization II. Contribution for the quantitative detection of circulating tumor DNA. Med Sci (Paris) 31(2):180–186
9. Foy CA, Parkes HC (2001) Emerging homogeneous DNA-based technologies in the clinical laboratory. Clin Chem 47(6):990–1000
10. Whitcombe D, Newton CR, Little S (1998) Advances in approaches to DNA-based diagnostics. Curr Opin Biotechnol 9(6):602–608
11. Carotenuto P, Roma C, Rachiglio AM et al (2010) Detection of KRAS mutations in colorectal carcinoma patients with an integrated PCR/sequencing and real-time PCR approach. Pharmacogenomics 11(8): 1169–1179
12. Dufort S, Richard MJ, de Fraipont F (2009) Pyrosequencing method to detect KRAS mutation in formalin-fixed and paran-embedded tumor tissues. Anal Biochem 391(2): 166–168
13. Ogino S, Kawasaki T, Brahmandam M et al (2005) Sensitive sequencing method for KRAS mutation detection by pyrosequencing. J Mol Diagn 7(3):413–421
14. Lievre A, Bachet JB, Valrie Boige V et al (2008) KRAS mutations as an independent prognostic factor in patients with advanced colorectal cancer treated with cetuximab. J Clin Oncol 26(3):374–379
15. Angulo B, Garca-Garca E, Martinez R et al (2010) A commercial real-time PCR kit provides greater sensitivity than direct sequencing to detect KRAS mutations: a morphology-based approach in colorectal carcinoma. J Mol Diagn 12(3):292–299
16. Tsiatis AC, Norris-Kirby A, Rich RG et al (2010) Comparison of sanger sequencing, pyrosequencing, and melting curve analysis for the detection of KRAS mutations: diagnostic and clinical implications. J Mol Diagn 12(4): 425–432
17. Sykes PJ, Neoh SH, Brisco MJ et al (1992) Quantitation of targets for PCR by use of limiting dilution. Biotechniques 13(3):444–449
18. Vogelstein B, Kinzler KW (1999) Digital PCR. Proc Natl Acad Sci U S A 96(16): 9236–9241
19. Diehl F, Schmidt K, Durkee KH et al (2008) Analysis of mutations in DNA isolated from plasma and stool of colorectal cancer patients. Gastroenterology 135(2):489–498
20. Dong SM, Traverso G, Johnson C et al (2001) Detecting colorectal cancer in stool with the use of multiple genetic targets. J Natl Cancer Inst 93(11):858–886
21. Sefrioui D, Sarafan-Vasseur N et al (2015) Clinical value of chip-based digital-PCR platform for the detection of circulating DNA in metastatic colorectal cancer. Dig Liver Dis 47(10):884–890
22. Shendure J, Porreca GJ, Reppas NB et al (2005) Accurate multiplex polony sequencing of an evolved bacterial genome. Science 309:1728–1732
23. Leamon JH, Lee WL, Tartaro KR et al (2003) A massively parallel picotiter plate based platform for discrete picoliter-scale polymerase chain reactions. Electrophoresis 24: 3769–3777
24. Dressman D, Yan H, Traverso G et al (2003) Transforming single DNA molecules into uorescent magnetic particles for detection and enumeration of genetic variations. Proc Natl Acad Sci U S A 100:8817–8822
25. Li M, Diehl F, Dressman D et al (2006) Beaming up for detection and quantification of rare sequence variants. Nat Methods 3(2): 95–97
26. Nakano M, Stone HA, Komatsu GPJ et al (2005) Single-molecule PCR using water-in-oil emulsion. J Biotechnol 102:117–124
27. Chen W, Balaj L et al (2013) BEAMing and droplet digital PCR analysis of mutant IDH1

mRNA in Glioma patient serum and cerebrospinal fluid extracellular vesicles. Mol Ther Nucleic Acids 2:e109
28. Pekin D, Skhiri Y, Baret J-C et al (2011) Quantitative and sensitive detection of rare mutations using droplet-based microfluidics. Lab Chip 11(13):2156–2166
29. Perkins G, Lu H, Garlan F, Taly V (2017) Droplet-Based Digital PCR: Application in Cancer Research. Adances in Clinical Chemistry (85): 44–91
30. Holtze C, Rowat AC, Agresti JJ et al (2008) Biocompatible surfactants for water-in- fluorocarbon emulsions. Lab Chip 8(10): 1632–1639
31. Lievre A, Bachet JB, Le Corre D et al (2006) KRAS mutation status is predictive of response to cetuximab therapy in colorectal cancer. Cancer Res 66(8):3992–3995
32. Xia YN, Whitesides GM (2006) Soft lithography. Angew. Chem. Int. Ed 37(5):551–575
33. Anna SL, Bontoux N, Stone HA (2003) Formation of dispersions using flow focusing in microchannels. Appl Phys Lett 82:364–366
34. Mazutis L, Baret J-C, Treacy P et al (2009) Multi-step microfluidic droplet processing: kinetic analysis of an in vitro translated enzyme. Lab Chip 9: 2902–2908
35. Soh J, Okumura N, Lockwood WW et al (2009) Oncogene mutations, copy number gains and mutant allele specific imbalance (MASI) frequently occur together in tumor cells. PLoS One 4(10):e7464
36. Mazutis L, Fallah Araghi A, Miller OJ et al (2009) Droplet-based microfluidic systems for high-throughput single DNA molecule isothermal amplification and analysis. Anal Chem 81(12):4813–4821

Part IV

Microfluidic Chips for Cell-Based Enrichment and Analysis

Chapter 13

Negative Enrichment of Circulating Tumor Cells in Blood Using a Microfluidic Chip

Hamizah A. Cognart and Chia-Pin Chang

Abstract

The enumeration and analysis of circulating tumor cells (CTCs) is an increasing interest for monitoring disease progression or response to treatment, specifically as a companion diagnostic for new anticancer drugs, and for research into the mechanisms of disease progression and metastases. Ideally, CTCs would be enriched from very small samples, with minimal handling, high recovery, and no requirement for the expression of specific surface markers. Here, we describe negative enrichment as the preferred approach for cancer cell isolation using a microfluidic platform.

Key words Circulating tumor cells (CTCs), Cancer cell isolation, Metastasis, Negative enrichment, White blood cell (WBC) depletion, Microfluidic

1 Introduction

Tumor cells detach from primary or metastatic sites and disseminate into the peripheral circulation. These cells are designated circulating tumor cells (CTCs) that attach to other regional and distant organs where they may proliferate and develop into eventual metastatic tumors. A CTC is an epithelial derived, nucleated, and often cytokeratin-positive cell that is not a normal constituent of blood. Its representation as a predictor of outcome in several types of cancer, including breast, prostate, colon, gastric, bladder, small and non-small cell lung carcinoma and melanoma, is associated with relatively short survival [1–9]. The extreme rarity of CTCs still makes them an elusive biomarker, particularly in early stages [10, 11]. Therefore, any assay for CTC enumeration requires the handling of a large number of normal cells.

In this protocol, we combine two technologies as previously described [12, 13] to isolate circulating tumor cells directly from whole blood using a microfluidic-based unbiased enrichment (negative selection) approach. The minimal sample handling permits high recovery of desired cells, allowing the detection of CTCs

Valérie Taly et al. (eds.), *Microchip Diagnostics: Methods and Protocols*, Methods in Molecular Biology, vol. 1547, DOI 10.1007/978-1-4939-6734-6_13, © Springer Science+Business Media LLC 2017

in much smaller samples than are currently used for clinical evaluation. CTCs are enriched without bias as to their surface antigen expression, and are not labeled with antibodies prior to detection. CTCs can be stained and visualized directly on the microfluidic chip. This unbiased enrichment approach could be used to assess the mutation status of CTC in real time.

2 Materials

Prepare and store all reagents at room temperature unless indicated otherwise. Diligently follow all waste disposal regulations when disposing waste materials. We do not add sodium azide to the reagents.

2.1 Micro Slit Membrane Components

A 10 μm thick micro slit membrane is designed to remove platelets and highly deformable red blood cells (RBCs) while retaining the majority of the other nucleated cells, including CTCs. The individual Parylene-based membrane is circular with a diameter of 13 mm when the active filter region is 9 mm in diameter. The membrane, shown in Fig. 1, consists of >100,000 precise slits with a slit dimension of 5.5 μm in width and 40 μm in length, which results in a fill factor of >40 % to reduce the pressure built up across the membrane to preserve the viability and morphology of the target cells. This micro slit membrane is packaged into a PMMA-based microfluidic chip.

2.2 Microfluidic Chip Components

Due to its characteristics of good light transition and high chemical resistance, Polymethyl methacrylate (PMMA) is a used as a substrate material to build a microfluidic chip to contain the micro slit membrane for CTC enrichment. The chip is designed as

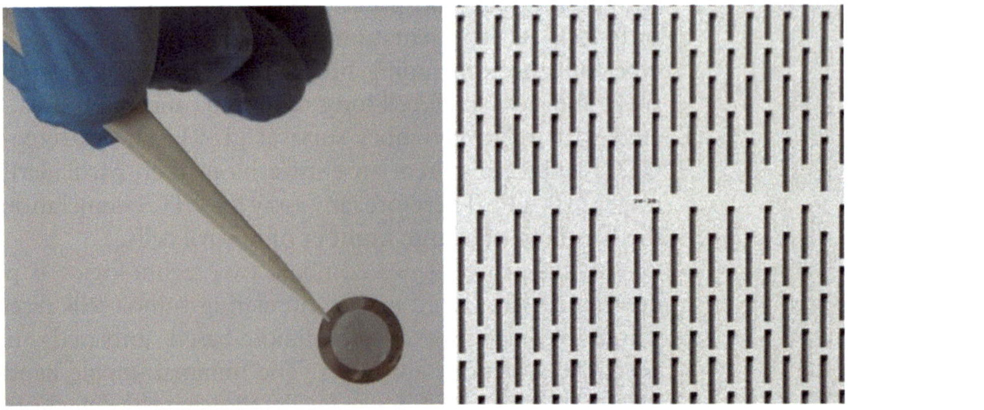

Fig. 1 Micro filter membrane of diameter 13 mm containing slits with dimension 5.5 × 40 μm and a fill factor of 44.5 %. *Left*: An overview picture of the membrane. *Right*: Image of micro slits taken at 300 × using scanning electron microscope (SEM)

25 mm wide and 75 mm long, which is the same size as the regular microscope slide to be easily mounted on the stage of microscope for inspection. The 3 mm thick chip is made with five layers of PMMA. Each PMMA layer is cut using CO_2 laser (5 mW, Universal, US) and then thermally bonded to each other. The Parylene filter membrane is mounted to the PMMA layers using silicone adhesive (RT 3140, Dow Corning, US). The microfluidic chip consists of an inlet, an outlet, a chamber to contain the Parylene micro filter membrane, and two guiding channels to bring the sample from inlet to the membrane, and then to the outlet. Both guiding channels are 0.5 mm in depth and 2 mm in width. Figure 2 shows the schematic of the microfluidic chip along with the steps in the cell isolation process.

2.3 Culture of Tumor Cell Lines

Human lung cancer cell line NCI—H1975 and human breast cancer cell line MCF—7 were purchased from American Type Culture Collection (Manassas, VA). NCI—H1975 cells were cultured in RPMI 1640 (Catlog #22400-089, Invitrogen) supplemented with 10 % Fetal Bovine Serum (FBS) (Catlog #04001-1A, Biological Industries). MCF—7 cells were cultured in DMEM (Catlog #11965-092, Invitrogen) supplemented with 10 % FBS. Cells were maintained at 37 °C in a humidified atmosphere containing 5 % CO_2 and harvested with trypsin before use. The cell suspensions were used only when their viability as assessed by Trypan blue exclusion exceeded 90 %.

2.4 Immunomagnetic Separation Components

1. Healthy donor's blood drawn into 6 mL BD Vacutainer K2 EDTA tubes (Becton Dickinson). Store at 4 °C.
2. Cellular preservative: Streck Cell Preservative™ (Streck).
3. EasySep™ Human CD45 Depletion Cocktail (StemCell Technologies, Canada). Store at 4 °C.
4. EasySep™ Magnetic Particles (StemCell Technologies, Canada). Store at 4 °C.
5. "The Big Easy" EasySep™ Magnet (StemCell Technologies, Canada).

2.5 Immunocytochemistry Components

1. Deionized water.
2. Phosphate-buffered saline (PBS; 1×): pH 7.4.
3. Blocking solution: 5 % bovine serum albumin (BSA) (Sigma) in 1× PBS. Store at 4 °C.
4. Fixing solution: 4 % paraformaldehyde (PFA) (Sigma) in 1× PBS.
5. 0.25 % Triton X – 100 (Sigma) in 1× PBS.

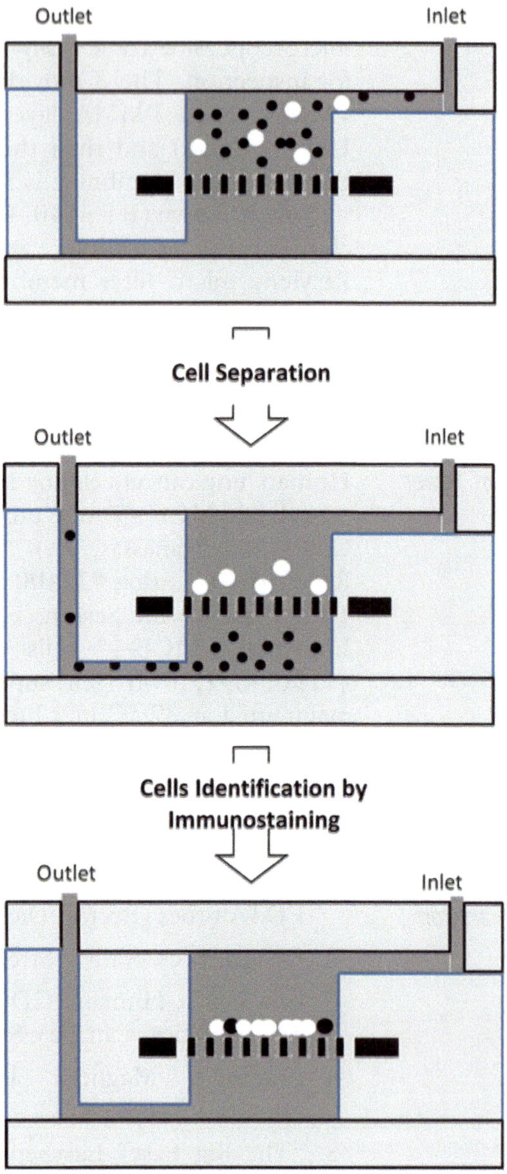

Fig. 2 Process of using the microfluidic chip. The micro slit membrane keeps the CTCs and nucleated cells on the membrane while depleting most of the RBCs and platelets. After cell separation, a series of immunostaining steps are carried out to differentiate cells isolated on the membrane. Design is not drawn to scale.

2.6 Antigen and Conjugates

1. Mouse anti-Human CD45 monoclonal antibody conjugated with Phycoerythrin(PE)-Dyomics 590 (Cedarlane Laboratories): 1 % in 500 μL of blocking solution.
2. Mouse anti-pan Cytokeratin monoclonal antibody conjugated with fluorescein isothiocyanate (FITC) (Abcam): 1 % in 500 μL of blocking solution.
3. Hoechst 33342 (Invitrogen): 0.2 % in 500 μL of 1× PBS.

3 Methods

Carry out all procedures at room temperature unless otherwise specified.

3.1 Sample Preparation

1. Add the cellular preservative to the drawn healthy donor's blood at a ratio of 1:1 (*see* **Note 1**).
2. Pipette 4 mL of preservative-added blood assay from the vacutainer tube into a 5 mL BD syringe barrel with a closed valve.
3. Spike the blood assay with the desired number of tumor cells (*see* **Note 2**).
4. Mix 100 μL of human CD45 depletion cocktail with 100 μL of magnetic particles in a microcentrifuge tube (*see* **Note 3**).
5. Add the antibody and particles mixture to the 5 mL BD syringe barrel containing the blood assay and incubate for 30 min.
6. Place the magnet around the barrel to separate the labeled WBCs.

3.2 Microfluidic Processing of Blood Samples Spiked with MCF—7/NCI—H1975 Cells

1. Retain the barrel within the magnet while placing the whole setup directly above an empty 5 mL syringe barrel connected to the inlet of the microfluidic chip (Fig. 3).
2. Open the valve of the syringe barrel containing the blood sample to release the WBC-depleted blood into the next empty syringe barrel under gravity. Discard the first syringe barrel.
3. Drive the WBC-depleted assay by a constant 3.5-mBar air pressure (MV20 pump, Ibidi) through the microfluidic chip for microfiltration process (*see* **Note 4**).

3.3 Staining and Enumeration of CTCs

1. After the microfiltration process of the blood assay, wash the chip with 2 mL of 1× PBS to remove the remaining RBCs on the micro slit membrane (*see* **Note 5**).
2. Fix the cells on the micro slit membrane with 500 μL of 4 % PFA and incubate for 15 min.
3. Rehydrate the cells by washing with 1 mL of 1× PBS.
4. Permeabilize the cells with 0.25 % Triton X – 100 and incubate for 10 min.
5. Wash the cells with 1 mL of 1× PBS.
6. Block the cells to avoid nonspecific binding sites with 500 μL of blocking solution and incubate for 30 min.
7. In a dark room, incubate the cells for 30 min with 500 μL of conjugated anti-Human CD45 and anti-pan Cytokeratin antibodies.
8. Wash the cells with 1 mL of 1× PBS.

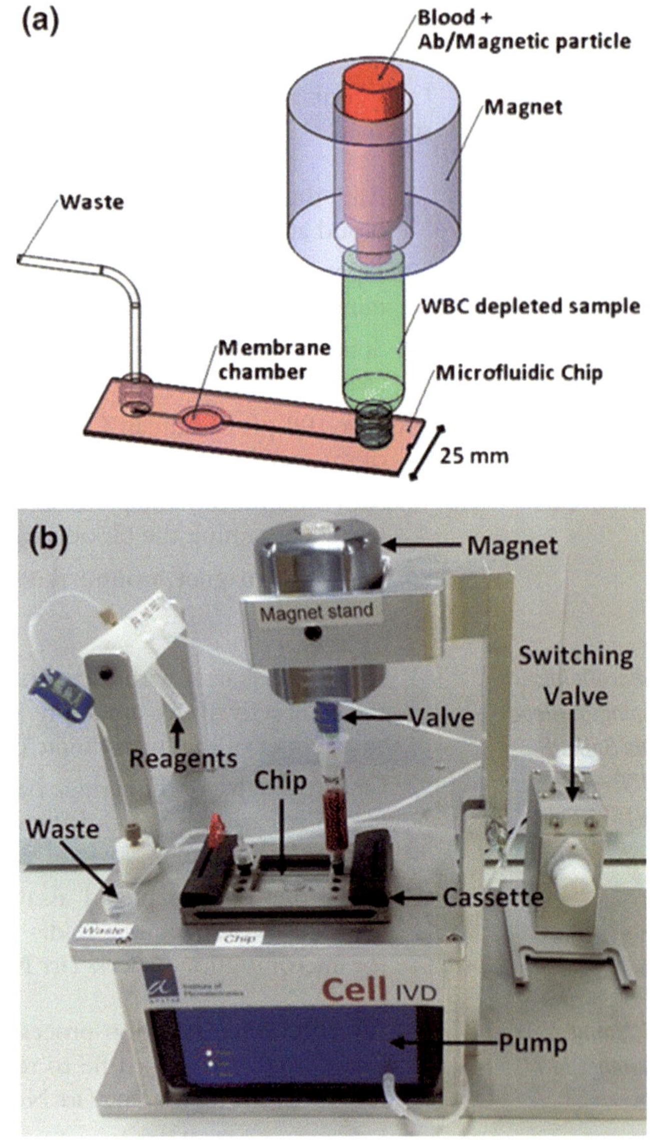

Fig. 3 Experimental setup for CTCs isolation from whole blood. (**a**) Schematic of main components in CTC isolation system and microfluidic chip setup. WBCs within the sample are depleted by immunomagnetic separation before flowing down to the chip for CTC isolation. (**b**) Photograph of the experimental setup for our CTC isolation system. Sample is driven by a pneumatic pressure regulated precisely by the pump (reproduced from [13] with permission from Springer).

9. Counter-stain with 500 μL of 0.2 % Hoechst 33342 for 2 min.
10. Wash the cells with 1 mL of 1× PBS.
11. Disconnect the chip from the connectors and mount chip on an upright fluorescent microscope for CTC enumeration and sample analysis (Fig. 4) (*see* **Note 6**).

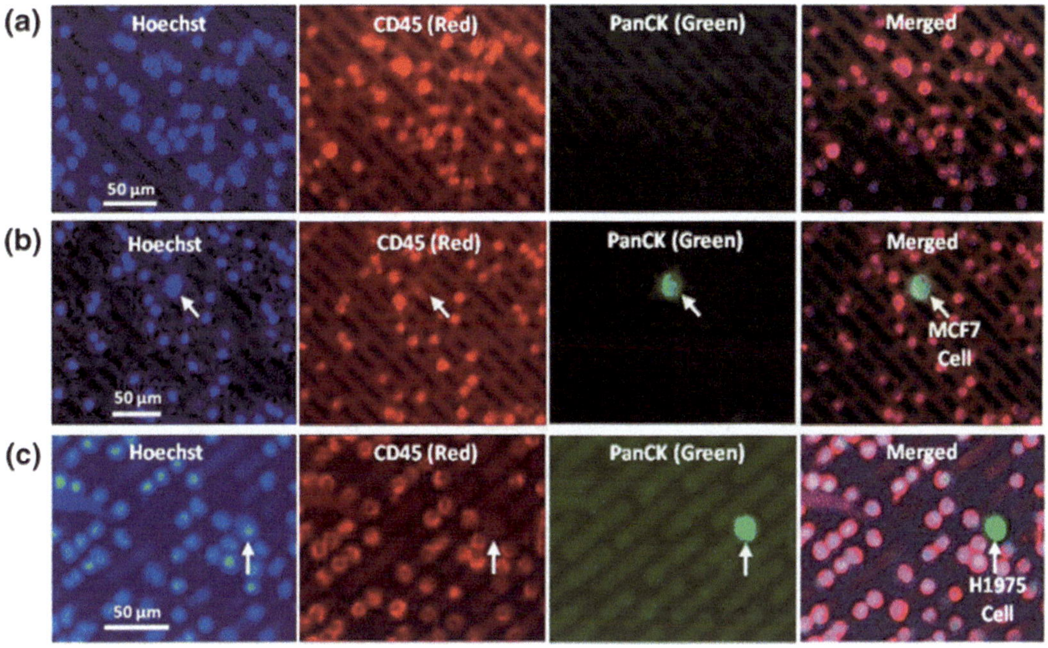

Fig. 4 Results of CTC isolation and enumeration. Images of cells captured on the micro slit membrane after processing healthy donor blood (**a**) negative control, blood without spiking of cancer cells. (**b**) Blood that was spiked with MCF—7 cells. (**c**) Blood that was spiked with NCI—H1975 cells. The cell nucleus were stained with Hoechst 33342 (*blue*), the WBCs were stained with PE—labeled anti-CD45 (*red*) and the cancer cells were stained by FITC—labeled antibodies Pan—Cytokeratin (*green*) (reproduced from [13] with permission from Springer).

4 Notes

1. Add the cellular preservative to the blood samples immediately after draw. Mix the sample thoroughly by inverting the tube five times to ensure homogeneity. Samples drawn were maintained in a 4 °C environment from the point of collection until the point of processing. Samples should be processed within 48 h from the time of draw.
2. Obtain concentration of cells by serial dilutions using 1× PBS. Use a hemocytometer to count the number of cells within a sample volume of 10 µL.
3. The dextran-coated magnetic microparticles used here are 1 µm diameter in size.
4. Leave behind a tiny volume of blood sample before stopping the air pressure to avoid the introduction of air bubbles.
5. At this point, the subsequent reagents should be introduced into the chip from the top of the syringe barrel and the same 3.5-mBar air pressure employed to deliver the reagents in all of the procedures, including washing, fixation, and staining steps.
6. We used a BX61 Olympus upright fluorescent microscope.

Acknowledgment

This work was supported by the Science and Engineering Research Council of A*STAR (Agency for Science, Technology and Research), Singapore under the grant number 1031490005.

References

1. Racila E, Euhus D, Weiss AJ et al (1998) Detection and characterization of carcinoma cells in the blood. Proc Natl Acad Sci U S A 95:4589–4594
2. Cristofanilli M, Budd G (2004) Circulating tumor cells, disease progression, and survival in metastatic breast cancer. Engl J Med 351:781–791
3. De Bono JS, Scher HI, Montgomery RB et al (2008) Circulating tumor cells predict survival benefit from treatment in metastatic castration-resistant prostate cancer. Clin Cancer Res 14:6302–6309
4. Cohen SJ, Punt CJ, Iannotti N et al (2008) Relationship of circulating tumor cells to tumor response, progression-free survival, and overall survival in patients with metastatic colorectal cancer. J Clin Oncol 26:3213–3221
5. Hiltermann TJN, Pore MM, Van Den Berg A et al (2005) Circulating tumor cells in small-cell lung cancer: a predictive and prognostic factor. Ann Oncol 23:2937–2942
6. Krebs MG, Sloane R, Priest L et al (2011) Evaluation and prognostic significance of circulating tumor cells in patients with non-small-cell lung cancer. J Clin Oncol 29: 1556–1563
7. Matsusaka S, Chin K, Oqura M et al (2010) Circulating tumor cells as a surrogate marker for determining response to chemotherapy in patients with advanced gastric cancer. Cancer Sci 101:1067–1071
8. Rao C, Bui T, Connelly M et al (2011) Circulating melanoma cells and survival in metastatic melanoma. Int J Oncol 38: 755–760
9. Gazzaniga P, Gradilone A, De Berardinis E et al (2012) Prognostic value of circulating tumor cells in nonmuscle invasive bladder cancer: a cellsearch analysis. Ann Oncol 23: 2352–2356
10. Allard WJ, Matera J, Miller MC et al (2004) Tumor cells circulate in the peripheral blood of all major carcinomas but not in healthy subjects or patients with nonmalignant diseases. Clin Cancer Res 10:6897–6904
11. Coumans FAW, Ligthart ST, Uhr JW et al (2012) Challenges in the Enumeration and Phenotyping of CTC. Clin Cancer Res 18:5711–5718
12. Sajay BG, Chang CP, Ahmad H et al (2013) Towards an optimal and unbiased approach for tumor cell isolation. Biomed Microdevices 15:699–709
13. Sajay BG, Chang CP, Ahmad H et al (2014) Microfluidic platform for negative enrichment of circulating tumor cells. Biomed Microdevices 16:537–548

Chapter 14

Microfluidic-Based Bacteria Isolation from Whole Blood for Diagnostics of Blood Stream Infection

Sergey Zelenin, Harisha Ramachandraiah, Asim Faridi, and Aman Russom

Abstract

Bacterial blood stream infection (BSI) potentially leads to life-threatening clinical conditions and medical emergencies such as severe sepsis, septic shock, and multi organ failure syndrome. Blood culturing is currently the gold standard for the identification of microorganisms and, although it has been automated over the decade, the process still requires 24–72 h to complete. This long turnaround time, especially for the identification of antimicrobial resistance, is driving the development of rapid molecular diagnostic methods. Rapid detection of microbial pathogens in blood related to bloodstream infections will allow the clinician to decide on or adjust the antimicrobial therapy potentially reducing the morbidity, mortality, and economic burden associated with BSI. For molecular-based methods, there is a lot to gain from an improved and straightforward method for isolation of bacteria from whole blood for downstream processing.

We describe a microfluidic-based sample-preparation approach that rapidly and selectively lyses all blood cells while it extracts intact bacteria for downstream analysis. Whole blood is exposed to a mild detergent, which lyses most blood cells, and then to osmotic shock using deionized water, which eliminates the remaining white blood cells. The recovered bacteria are 100 % viable, which opens up possibilities for performing drug susceptibility tests and for nucleic-acid-based molecular identification.

Key words Sepsis, Blood stream infection, Microfluidics, Diagnostic, Pathogens, Bacteremia, Point-of-care diagnostics

1 Introduction

Bloodstream infections (BSI) are associated with high morbidity and mortality rates. Without appropriate therapy, BSI may lead to severe sepsis, septic shock, and multiorgan failure syndrome. In the prospective cohort study published by Levy et al, it was discovered that the hospital mortality of patients with BSI is in range from 28 to 41 % [1]. It has been shown that the survival of infected patients is directly correlated to the time before administration of appropriate antibiotics [2]. Therefore, early appropriate initiation of adequate antimicrobial therapy is one of the most important factors for the outcome of the disease. Despite the

Valérie Taly et al. (eds.), *Microchip Diagnostics: Methods and Protocols*, Methods in Molecular Biology, vol. 1547,
DOI 10.1007/978-1-4939-6734-6_14, © Springer Science+Business Media LLC 2017

urgent need for rapid diagnostic tests, traditional blood cultures are still considered the gold standard of BSI diagnosis. The turnaround time for blood culture is lengthy, ranging from 24 to 72 h. This long turnaround time, especially for the identification of antimicrobial resistance, is driving the development of molecular diagnostics, often based on PCR. These methods are used to detect pathogens either from blood culture bottles [3–5] or directly from blood [6–11]. However, the need for complex, multistep sample preparation remains the bottleneck for the further development of molecular diagnostics in clinical settings. Main challenge in pathogen purification from septic blood related to very low concentration of bacteria which is often from ≤1 to 10 CFU per mL in adult blood and above 100 CFU per mL in infants and young children [12–14].

Here, we describe a microfluidic-based sample preparation assay to isolate and enrich bacteria from complex sample matrices such as blood. The bacteria isolation approach relies on inherent properties of cells to regulate cell volume/integrity—enabling controlled blood cell lysis while intact bacteria are readily recovered and separated [15]. Because bacteria have more rigid cell walls than blood cells [16–18], this method isolates bacteria in a two-step operation: a mild detergent, saponin, lyses most of the blood cells, and then deionized water provides an osmotic shock to lyse the few remaining white blood cells (WBCs). We confirm that bacteria subjected to this process in blood are viable and readily recovered for downstream processing. The microfluidic device is fabricated by casting polydimethylsiloxane (PDMS) polymer on a resist-structured silicon wafer according to standard soft lithographic techniques. The experimental setup for the microfluidic technique consists of the microfluidic device and three syringe pumps (Fig. 1). Staggered herringbone structure inside microchip channels was used for efficient mixing of liquids [19]. The device is first primed by filling with 1×PBS and connected to respective syringes without trapping air bubbles. The first syringe pump drives the whole blood through inlet 1 at 24 μL/min, and the second pump drives a lysis solution containing saponin through inlet 2 at 24 μL/min while the third pump flows deionized water through inlet 3 at 48 μL/min. The output of the device is 100 % viable bacteria while all blood cells are successively lysed. Bacteria collected from patient-blood samples can be analyzed by two complementary methods: molecular-based testing and antibiotic-susceptibility testing. Genetic analysis of bacteria gives information about strains that cause BSI symptoms and about potential antibiotic treatment. Here, we described in detail the whole procedure including manufacture of microchips, selective cell lysis protocol itself, and validation of recovered bacteria and blood cell debris in lysates.

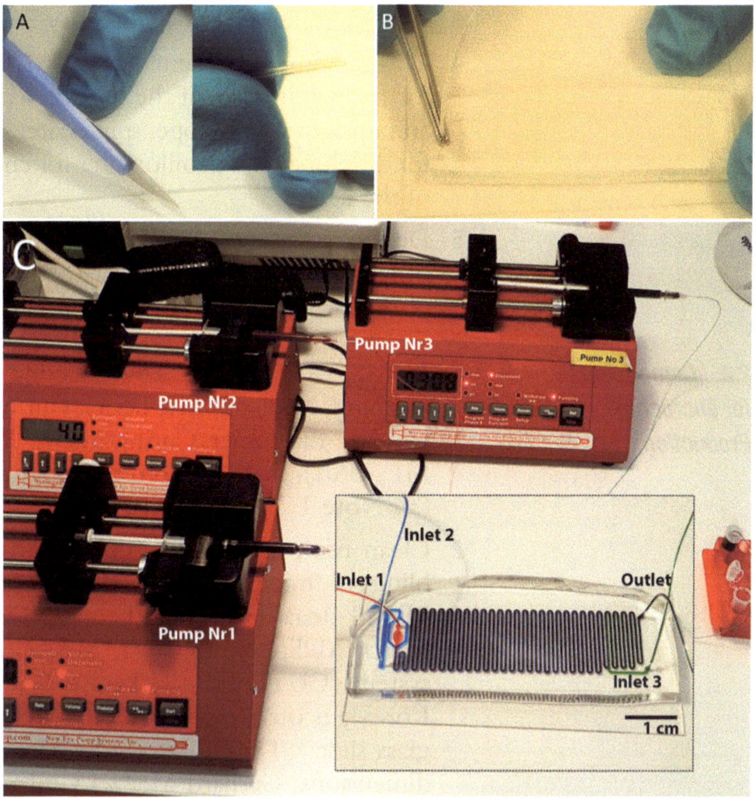

Fig. 1 Microfluidic device for selective blood cell lysis. (**a**–**b**) Assembly of the microfluidic device. Connecting tube tip should be angle cut to make it easy to inert the tube inside the PDMS hole and avoid shortcutting the fluid during pumping (**a**). The tubing should be inserted properly using tweezers (**b**). The hole is slightly smaller than the tubing diameter. (**c**) Final setup. The setup includes three syringe pumps connected to the microchip (magnified view) and collecting tubes for initial microchip priming and collection of final blood lysate

2 Materials

Prepare all the solutions using deionized water. Blood samples from patients with sepsis should be used the same day. The control blood sample from healthy donors could be kept at +4 °C up to 48 h before use, but is recommended to use fresh blood when possible.

2.1 Components for Master Mold Production

1. SU-8 Negative Epoxy Series Resists and developer solution from Microchem (SU-8 2000 Series, Microchem, Newton, MA, USA).
2. Developer mr-Dev600 (Microresist technology, Berlin, Germany), acetone, and ethanol (Sigma-Aldrich, Sweden).

3. Silicone wafer <100> (University Wafer, Boston, USA).
4. Clean room equipment—standard MEMS cleanroom is used for the production of the master: Mask aligner with UV lamp and upright microscope, spin coater, hot plates, and wet bench for developer and profilometer for checking the channel height of the device.
5. Design software (AutoCAD, L-Edit, Adobe illustrator, or other CAD software) to draw the pattern of design that will be printed on transparent photomask.

2.2 Components for Microchip Production

1. PDMS (SYLGARD, 184 Silicone Elastomer Base, DOW Corning, Midland, MI, USA), Curing agent (SYLGARD, 184 Silicone Elastomer Curing agent, DOW Corning, Midland, MI , USA), store at room temperature in well-ventilated place (*see* **Note 1**).
2. Clean room equipment and material for casting and assembling of microchips: chemical resistant vacuum pump for PDMS degasification (Chemical Duty Pump, WP6122050, Merck Millipore, Billerica, MA, USA), low pressure plasma system (CUTE-MP/R, Femto Science, Inc., Suwon, South Korea), Scotch tape (Scotch Magic Tape 810), microscope glass slides (Thermo Scientific Menzel-Gläser, 100 OTG 90° dimensions, Gerhard Menzel GmbH, Germany), Harris Unicore biopsy punch and plunger (Harris Uni-Core Punch, 0.75 mm, Sigma Aldrich, Sweden), Cutting Mat Single Side, Tygon tubes (microbore tubing, 0.010″ × 0.030″OD, Cole-Parmer, Vernon Hills, IL, USA).
3. Albumin fraction V from bovine serum (VWR, Belgium) (*see* **Note 2**).

2.3 Bacteria Culture and Selective Cell Lysis Components

1. Agar plates: Weigh 20 g Blood agar base (1.10886.0500, Merck, Darmstadt, Germany), transfer to 500 mL glass bottle with screw cap, add 500 mL of deionized water, autoclave (15 min at 121 °C), cool to 50 °C–60 °C, pour 12–15 mL of agar on 10 cm Petri plate, leave to solidify agar completely, store at +4 °C (*see* **Note 3**).
2. Saponin solution: Weigh 5 g saponin from Quillaja Bark pure DAB (AppliChem, Darmstadt, Germany), transfer to 50 mL Falcon tube, add 40 mL of deionized water, close tube and mix on rotating table till complete dissolving of saponin, adjust volume till 50 mL, store in small aliquots at −20 °C (*see* **Note 4**).
3. DPBS/Modified (HyClone Laboratories, Utah, US), bacteria BL21-AI™ One Shot® Chemically Competent *E. coli* (Thermofisher, Stockholm, Sweden), syringe pump (NE-1000, New Era Pump Systems, Farmingdale, NY, USA).

2.4 FACS Components

1. Antibody T cell multicolor marker panel (Abcam, Cambridge, UK).
2. Flow count beads (Beckman coulter, Krefeld, Germany).
3. Flow cytometer (Gallios Flow Cytometer, Beckman Coulter).
4. Kaluza analysis software (Beckman Coulter, Krefeld, Germany).

3 Methods

3.1 Fabrication of Master Mold

Two-layer structures in SU-8 resist are produced according to the supplier's recipe (SU-8 2000 Series, Microchem, Newton, MA, USA) using standard MEMS technology. A normal process follows the following steps: spin coat, soft bake, expose, post expose bake, and develop. A detailed protocol can be found at the following web: http://www.microchem.com/pdf/SU8_50-100.pdf

For the first layer (150 μm thick) follow the instruction for the polymer SU-8 100 and for the second layer (25 μm) use the SU8-25

1. Spin-coat the silicone substrate with a negative photoresist (SU-8 100 for the first layer). Start with dispensing approximate 3–4 ml of SU-8 at the center of a 4-inch wafer and let it spread. For spin-coating, program the spin-coater in a two step process: (a) Ramp to 500 rpm at 100 rpm/sec acceleration and hold this speed for 5 s to allow the resist to cover the entire surface, following by (b) Ramp to final spin speed of 1750 rpm at an acceleration of 300 rpm/sec and hold for a total of 30 s. The final speed (1750–2000 rpm) given approximately 150 μm thick resist for SU-8 100.
2. Bake the photoresist-coated silicon wafer on a hot plate at 65 and 95 °C for 20 and 50 min respectively.
3. Expose photoresist-coated silicon wafer through a photomask (mask nr 1 = first layer, defining the channels), expose for a time that gives approx. energy of 600 mJ/cm^{2}s.
4. Post bake the exposed wafer on a hot plate at 65 °C and 95 °C for 1 and 12 min respectively.
5. Develop the photoresist for 15 min in developer solution.
6. Rinse with isopropyl alcohol, then apply gentle stream of air or nitrogen to dry the wafer.
7. Put the wafer hot plate at 65 °C for 3 min to completely dry the wafer.
8. For second layer preparation follow the **steps 1–5** for the polymer SU-8 25 (*see* **Note 5**).

9. After microscope inspection, a final hard bake is recommended: UV expose the wafer using through a transparent glass mask for 60 s, and bake on plate at 65 °C and 95 °C for 3 and 7 min respectively.
10. Check the height of the two layers using a profilometer.

3.2 Fabrication of PDMS Microfluidic Device

Prepare before experiment: Petri dish with silicon wafer for manufacturing of microchips, vacuum pump, desiccator, and oven. Carry out all procedures in clean dust-free room till BSA treatment.

1. Place the silicon-SU 8 master mold with microchip structures facing up in the thermostable Petri dish (diameter 15 cm), attach the master mold to the Petri dish surface by tape.
2. Weigh 40 g PDMS and 4 g of curing agent directly in a plastic disposable beaker (*see* **Note 6**), mix thoroughly using plastic spoon or plastic rod, pour PDMS mixture onto the master mold in the Petri dish, place it in the desiccator to remove the air bubbles.
3. Apply vacuum to remove air bubbles from PDMS mixture [6].
4. Transfer the Petri dish with degassed PDMS mixture in the oven and keep at 65 °C for at least 5–6 h. A transparent elastomeric PDMS solid will be formed. The master mold with PDMS can now be kept at room temperature until further use.
5. Prepare clean place and cut out PDMS slab using scalpel with all microchip structures in one piece (*see* **Note 7**).
6. Take away slab from Petri dish (*see* **Note 8**), place on clean tray and cut the PDMS slab into individual microchips, transfer to 9 cm Petri dish, and keep in clean place at room temperature.
7. Place PDMS microchips on clean tray for PDMS cutting, use biopsy puncher to make inlet and outlet holes (*see* **Note 9**), clean PDMS fragment and glass with tape (*see* **Note 10**).
8. Place cleaned glasses and PDMS fragments with the features facing up inside the oxygen plasma chamber.
9. Set the Oxygen pressure to 5 mbar and flow rate to 10 sccm, and set the power to 50 watt.
10. Perform plasma treatment for 30 s.
11. After plasma treatment, carefully place the PDMS and the glass substrate in contact and remove the trapped air, simply by pressing with the tweezers.
12. Transfer and keep the glass with PDMS microchip for at least 10 min in an oven at 110 °C for irreversibly bonding of the microchip. Alternatively, a standard hotplate (80 °C for 10 min) can be used for bonding. Take out the microchip, wait for few min until the temperature reduced to room temperature, and

cover both sides of microchip with tape to prevent dust entering.

13. Angle cut three Tygon tubes by sharp blade with length of each tube 10–15 cm. Attach tubes using tweezers to inlet 1, inlet 3, and outlet of microchip. Place free ends of Tygon tubes into the waste vessel.
14. Fill up syringe with 1 % BSA solution; connect one end of Tygon tube of length 15–20 cm to syringe and second end to inlet 2 of microchip. Fill up microchip with BSA solution by pumping with moderate flow (10–100 μL/min). Remove air bubbles trapped in micro channels by gentle tapping or pressing of microchip surface with tweezers.
15. Cover the microchip with tape and keep it overnight at 4 °C.

3.3 Selective Blood Cell Lysis

Carry out all procedures at room temperature unless otherwise specified. Prepare before experiment: blood samples, bacteria suspension, three syringe pumps, blunt round end tweezers, vessels for waste and lysate, four 1 mL syringes with blunt end needles and connecting tubes.

1. Prepare blood sample (positive control sample) by mixing 1 mL of blood sample from healthy donor with10 μL of bacteria suspension in 1× PBS (10^5 CFU per mL, *see* **Note 11**).
2. Attach connecting tubes to inlet 1, inlet 3, and outlet of microchip (*see* **Note 12**). Place free ends of tubes into vessel for waste. Fill up syringe with 1× PBS, connect one end of Tygon tube to syringe and second end to inlet 2 of microchip. Replace BSA solution by flowing of PBS with moderate flow (10–100 μL/min) through the microchip. Remove air bubble trapped in micro channels by gentle tapping or pressing of microchip surface with tweezers.
3. Fill up three 1 mL syringes with (a) 1 mL of blood sample, (b) 1 mL of 2 % saponin, and (c) 1 mL of deionized water, attach the needles and remove air from syringes (*see* **Note 13**). Attach connecting Tygon tubes to syringes. Use the length of connecting tube sufficient to reach microchip from syringe pumps. Push out air from tubes and connect syringes to pumps.
4. Place microchip on stand with syringe pumps. Connect tube of syringe with blood sample to inlet 1 of microchip, tube of syringe with 2 % saponin to inlet 2, and tube of syringe with water to inlet 3. Place free end of tube connected to outlet to the waste vessel.
5. Select injection mode on all pumps. Set flow rate 24 μL/min for blood and saponin and 48 μL/min on pump with water.
6. Start saponin and blood injection first and 30 s later the third pump with water.

7. Run the sample and solutions for 1 min to stabilize the flow conditions. Replace tube from waste vessel to vessel for lysate collection. Analyze concentration of bacteria in the blood positive control input and in the lysate (*see* **Note 14**). Analyze the efficiency of blood cell lysis and size of cell debris using FACS machine (*see* **Note 15**).

4 Notes

1. PDMS and Curing agent are highly viscous liquids. Ration between PDMS and Curing agent is important to get the same quality of PDMS slab. Making aliquot of Curing agent in dispenser bottle helps to get right weight of Curing agent.
2. Weigh 1 g BSA, transfer to 200 mL beaker, add 99 mL of PBS, mix, sterilize by passing through the nitrocellulose filter with pore size 0.2 μm, store at +4 °C.
3. Close 500 mL bottle with water and agar, shake it quickly by turning upside down few times to disperse agar clamps, open cap (one half turn) before sterilization. Close bottle completely after sterilization, vortex by hands to make homogeneous solution. Petri plates could be placed in piles (10–20 per pile) for agar solidification to save bench space. Place agar plates in plastic bag, store at +4 °C with agar site up for 2–4 weeks. Optionally, to extend agar storage time, bottle with sterilized agar could be closed, cool to room temperature, and store at +4 °C. Use microwave oven (30 % of maximal power) to melt agar in bottle before preparing of agar plates. It is no need for additional sterilization.
4. Shaking of saponin solution during solubilization leads to bulky foam formation. Leave saponin solution standing until most of the foam has disappeared. Adjust volume by adding deionized water and mix solution. Sterilize solution by passing through the nitrocellulose filter with pore size 0.2 μm. Make aliquots and keep saponin frozen at −20 °C. Saponin solution could be kept at room temperature for few days before use. Do not use PBS for dissolve saponin because the solution can start to precipitate.
5. For two layer SU-8 fabrication, there are two ways. In the first version (described above), you process the first layer first all the way to development and then the second layer. Before applying the polymer for the second layer, sometimes adhesion promoter such as HMDS or a short plasma treatment might be necessary to promote good adhesion of polymer solution. An alternative method is to first expose the first layer, followed by postexposure bake. After this, spin-coat the second layer without the development step and do soft and hard bake followed

by alignment using the microscope of the mask aligner and UV expose the second layer. Bake postexposure and then develop both layers at the same time. If it is difficult to see through the SU-8 layer to the alignment markers, a tip is to carefully add developing solvent during the soft bake using 10 μL pipette tips. The developer opens up the aligning marks without spreading over the wafer.

6. To avoid tearing while peeling of PDMS after cutting, for the first time, use 40 g PDMS and 4 g of curing agent to make thick PDMS.
7. Place Petri dish with PDMS mixture in desiccator, connect to vacuum pump, and apply vacuum to remove bubbles in the PDMS mixture. To facilitate collapsing of bubble, after 5–10 min of vacuum, disconnect vacuum pump from desiccator by turning of valve, open carefully desiccator to equalize pressure, and apply vacuum again. Continue procedure until no bubbles are visible (approximately 30 min).
8. To avoid particles, use gloves and lab coats; clean the surface using dust-free paper with isopropanol; open silicon wafer with PDMS and cut PDMS slab only in a clean cabinet. Care should be taken while cutting the PDMS, for instance one can hold the scalpel at an angel of 45°. Punch PDMS with scalpel as much as possible outside microchip structures, cut out all microchips in one piece by continuous moving of scalpel to prevent forming of small pieces of PDMS.
9. Make punches straight down. Wipe the puncher clean after each punch. Transfer PDMS fragments in new Petri dishes. Clean puncher and plastic board using dust-free paper with isopropanol.
10. Surface of microscope glass and PDMS fragment should be dust free before assembling of complete microchip structure. Put PDMS fragment with individual microchip channel on the table with microstructure side up. Clean the front side of microchip by covering with tape and removing it. Cover front side of microchip with tape until plasma treatment. Turn slab around and clean backside of PDMS once with tape. Clean a glass side one with tape, place new tape until plasma treatment.
11. The bacteria suspension for control blood sample experiments must be prepared fresh and used within 2 days. To prepare bacteria suspension, use blood agar plate (diameter ≈ 10 cm) with overnight bacteria culture with density about 10^3 CFU per plate. Add 5 mL of PBS to agar plate and detach bacteria from plate by shaking on rotation table for 10 min with rotation speed ≈100 rpm. Pull down bacteria from suspension by centrifugation at room temperature for 4 min at 5000 × *g*. Resuspend bacteria pellet in PBS and adjust optical density of

bacteria suspension until $OD_{600} = 0.5$ by PBS. Analyze concentration of bacteria in final suspension by plating on agar plates and counting of colonies after overnight incubation. Calculate exact concentration of bacteria in the 0.5 stock suspension. The concentration of bacteria is normally between 10^7 and 10^8 CFU per mL. Calculate dilution coefficient and adjust bacteria concentration by dilution in PBS until 10^6 CFU per mL. Mix 50 μL of 10^6 CFU per mL bacteria suspension and 450 μL of PBS to prepare bacteria suspension before blood spiking. Vortex for 3–5 s bacteria suspension before taking of each aliquot and increase vortex time until 1–2 min if bacteria were staying for long time.

12. Sharp blade should be used for cutting of connecting tube with gentle pressing to keep open of the tube end. The cut should not be straight with optimal angle around 45 °C (Fig. 1a). Use tweezers to attach connecting tubes to the inlet and outlet holes of the microchip. Control connection of tube to the microchip by slow racing of microchip via connecting tube (Fig. 1b).
13. Care should be taken to avoid air bubbles while connecting the syringes to pumps and to the microchip.
14. The viability of the recovered bacteria is analyzed by plating (Fig. 2). Take 100 μL aliquot of donor blood spiked with bacteria (10^3 CFU per mL of blood) before selective cell lysis experiment, add to 900 μL of PBS, save it at +4 °C, and use as the input control. Run selective cell lysis experiment, collect lysate, and mix 100 μL of blood cell lysate with 900 μL of

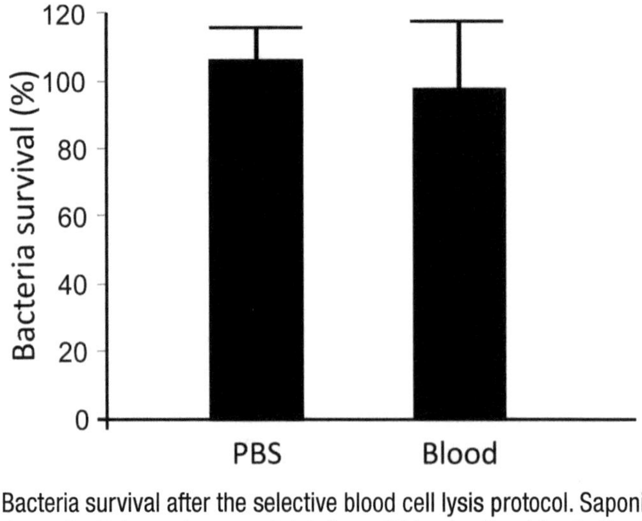

Fig. 2 Bacteria survival after the selective blood cell lysis protocol. Saponin concentrations of 1 % is used to completely lyse all blood cells without affecting the viability of the bacteria

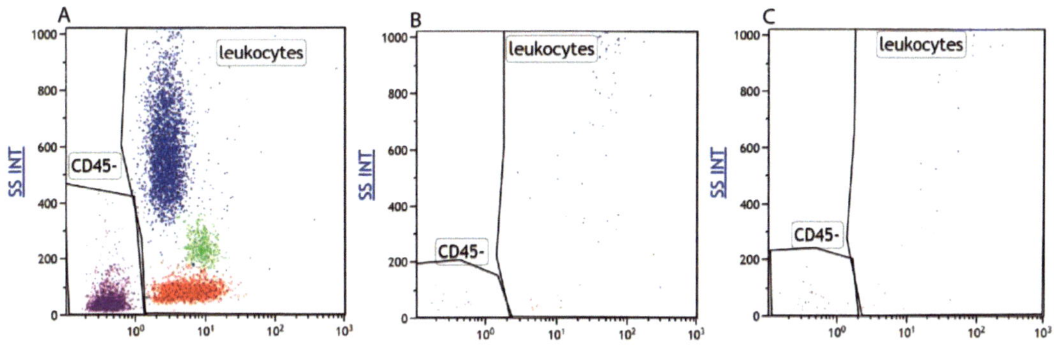

Fig. 3 FACS analysis of the effect of the detergent on the size of blood debris and amount of remaining blood cells in lysate. (**a**) The osmotic shock without saponin treatment only slightly decreases amount of white blood cells in the lysates. (**b**) Incubation of the blood sample with 0.5 % saponin prior to the osmotic shock dramatically decreases the amount of white blood cells. (**c**) Only traces of white blood cells or cell debris are observed for lysis using saponin concentration of 1 %

PBS. Label blood agar plates and load equal amount of sterile glass beads (3 mm diameter) on agar by sterile tea spoon. Vortex the input control and diluted blood cell lysate, transfer 100 μL of each sample on corresponding agar plate, close lead and spread bacteria suspension by shaking beads for 30–60 s on agar (avoid round movements). Discard beads in biological waste; incubate agar plates overnight at 37 °C. Gel Documentation System (Bio-Rad) with colony counting settings can be used for scanning of plates and analysis of CFU on plates. Alternatively, one can use manual counting of colonies.

15. Analysis of blood cells and debris using FACS machine (Fig. 3). Aliquot 500 μL of blood lysate in 1.5 mL Eppendorf tube, centrifuge at room temperature for 5 min at 180 g, resuspend the pallet in 100 μL of 1× PBS. Add 5 μL of antibody cocktail and incubate for 15 min at room temperature. Add 20 μL of flow count beads (Beckman coulter) to calculate the absolute number of cells. Place the solution into the flow cytometer tubes and place them in the flow cytometer carousel. Flow cytometer results were analyzed using Kaluza software.

Acknowledgment

This work was supported by the European Commission (projects FP7 InTopSens and IMI RAPP-ID)

References

1. Levy MM, Artigas A, Phillips GS, Rhodes A, Beale R, Osborn T et al (2012) Outcomes of the surviving sepsis campaign in intensive care units in the USA and Europe: a prospective cohort study. Lancet Infect Dis 12(12):919–924
2. Kumar A, Roberts D, Wood KE, Light B, Parrillo JE, Sharma S et al (2006) Duration of hypotension before initiation of effective antimicrobial therapy is the critical determinant of survival in human septic shock. Crit Care Med 34(6):1589–1596
3. Tissari P, Zumla A, Tarkka E, Mero S, Savolainen L, Vaara M et al (2010) Accurate and rapid identification of bacterial species from positive blood cultures with a DNA-based microarray platform: an observational study. Lancet 375(9710):224–230
4. Turenne CY, Witwicki E, Hoban DJ, Karlowsky JA, Kabani AM (2000) Rapid identification of bacteria from positive blood cultures by fluorescence-based PCR-single-strand conformation polymorphism analysis of the 16S rRNA gene. J Clin Microbiol 38(2):513–520
5. Wellinghausen N, Wirths B, Essig A, Wassill L (2004) Evaluation of the hyplex bloodscreen multiplex PCR-enzyme-linked immunosorbent assay system for direct identification of gram-positive cocci and gram-negative bacilli from positive blood cultures. J Clin Microbiol 42(7):3147–3152
6. McCann CD, Jordan JA (2014) Evaluation of MolYsis™ Complete5 DNA extraction method for detecting Staphylococcus aureus DNA from whole blood in a sepsis model using PCR/pyrosequencing. J Microbiol Methods 99:1–7
7. Vince A, Lepej SZ, Barsić B, Dusek D, Mitrović Z, Serventi-Seiwerth R et al (2008) LightCycler SeptiFast assay as a tool for the rapid diagnosis of sepsis in patients during antimicrobial therapy. J Med Microbiol 257(Pt 10):1306–1307
8. Tang Y-W, Kilic A, Yang Q, McAllister SK, Li H, Miller RS et al (2007) StaphPlex system for rapid and simultaneous identification of antibiotic resistance determinants and Panton-Valentine leukocidin detection of staphylococci from positive blood cultures. J Clin Microbiol 45(6):1867–1873
9. Jordan JA, Durso MB (2005) Real-time polymerase chain reaction for detecting bacterial DNA directly from blood of neonates being evaluated for sepsis. J Mol Diagn 7(5):575–581
10. Wellinghausen N, Kochem A-J, Disqué C, Mühl H, Gebert S, Winter J et al (2009) Diagnosis of bacteremia in whole-blood samples by use of a commercial universal 16S rRNA gene-based PCR and sequence analysis. J Clin Microbiol 47(9):2759–2765
11. Weller SA, Cox V, Essex-Lopresti A, Hartley MG, Parsons TM, Rachwal PA et al (2012) Evaluation of two multiplex real-time PCR screening capabilities for the detection of Bacillus anthracis, Francisella tularensis and Yersinia pestis in blood samples generated from murine infection models. J Med Microbiol 61(Pt 11):1546–1555
12. Kellogg JA, Manzella JP, Bankert DA (2000) Frequency of low-level bacteremia in children from birth to fifteen years of age. J Clin Microbiol 38(6):2181–2185
13. Henry NK, McLimans CA, Wright AJ, Thompson RL, Wilson WR, Washington JA (1983) Microbiological and clinical evaluation of the isolator lysis-centrifugation blood culture tube. J Clin Microbiol 17(5):864–869
14. Yagupsky P, Nolte FS (1990) Quantitative aspects of septicemia. Clin Microbiol Rev 3(3):269–279
15. Zelenin S, Hansson J, Ardabili S, Ramachandraiah H, Brismar H, Russom A (2015) Microfluidic-based isolation of bacteria from whole blood for sepsis diagnostics. Biotechnol Lett 37(4):825–830
16. Pichereau V, Hartke A, Auffray Y (2000) Starvation and osmotic stress induced multiresistances. Influence of extracellular compounds. Int J Food Microbiol 55(1–3):19–25
17. Corcoran BM, Stanton C, Fitzgerald G, Ross RP (2008) Life under stress: the probiotic stress response and how it may be manipulated. Curr Pharm Des 14(14):1382–1399
18. Ho SN (2006) Intracellular water homeostasis and the mammalian cellular osmotic stress response. J Cell Physiol 206(1):9–15
19. Stroock AD, Dertinger SKW, Ajdari A, Mezic I, Stone HA, Whitesides GM (2002) Chaotic mixer for microchannels. Science 295(5555):647–651

Chapter 15

Microfluidic Platform for Parallel Single Cell Analysis for Diagnostic Applications

Séverine Le Gac

Abstract

Cell populations are heterogeneous: they can comprise different cell types or even cells at different stages of the cell cycle and/or of biological processes. Furthermore, molecular processes taking place in cells are stochastic in nature. Therefore, cellular analysis must be brought down to the single cell level to get useful insight into biological processes, and to access essential molecular information that would be lost when using a cell population analysis approach. Furthermore, to fully characterize a cell population, ideally, information both at the single cell level and on the whole cell population is required, which calls for analyzing each individual cell in a population in a parallel manner. This single cell level analysis approach is particularly important for diagnostic applications to unravel molecular perturbations at the onset of a disease, to identify biomarkers, and for personalized medicine, not only because of the heterogeneity of the cell sample, but also due to the availability of a reduced amount of cells, or even unique cells. This chapter presents a versatile platform meant for the parallel analysis of individual cells, with a particular focus on diagnostic applications and the analysis of cancer cells. We first describe one essential step of this parallel single cell analysis protocol, which is the trapping of individual cells in dedicated structures. Following this, we report different steps of a whole analytical process, including on-chip cell staining and imaging, cell membrane permeabilization and/or lysis using either chemical or physical means, and retrieval of the cell molecular content in dedicated channels for further analysis. This series of experiments illustrates the versatility of the herein-presented platform and its suitability for various analysis schemes and different analytical purposes.

Key words Single cell, Parallelization, Microfluidics, Circulating tumor cells, Cell trapping, Cell permeabilization

1 Introduction

Cell analysis is conventionally conducted at the scale of a population, consisting of a million to a billion of cells, yielding average data at the population level. It is however now well acknowledged that cell populations and tissues are heterogeneous [1]. Tissues, for instance, comprise different cells types. Furthermore, in any cell population, cells are asynchronous for cell cycles and biological processes such as, e.g., apoptosis and cell differentiation.

Valérie Taly et al. (eds.), *Microchip Diagnostics: Methods and Protocols*, Methods in Molecular Biology, vol. 1547,
DOI 10.1007/978-1-4939-6734-6_15, © Springer Science+Business Media LLC 2017

Finally, rare cells or stochastic events occurring in individual cells often bear essential information for the understanding of fundamental processes. Consequently, in general, cellular analysis must be brought down to the single cell level to get useful insight into biological processes, and not to miss the information, which would be lost when using an analytical approach at the level of the whole cell population.

Analyzing cells at the single cell level is also highly relevant for diagnostic applications, not only when heterogeneous populations are encountered but also when only a handful of cells or even individual cells are available for analysis. First, as for any tissue, tumors comprise various cell types including cancer cells and cancer stem cells (CSCs) as well as stromal cells such as fibroblasts [2]. Only a small portion of the cells (<20 %) does bear the hallmarks of the disease [2], which become masked using a population analysis approach. Therefore, all cells must be analyzed separately to unravel molecular perturbations at the origin of the disease, to identify biomarkers and for the development of personalized medicine [3]. Similarly, circulating tumor cells (CTCs), found in liquid biopsies, are highly diluted among a large amount of blood cells (1–10 CTCs/mL blood *vs.* >10^9 blood cells), and they exhibit extremely different phenotypes [4], which calls out for a single cell analysis approach for their molecular characterization, after their selective isolation. Ideally, in these cases, information at both the single cell and population levels is required to actually fully characterize heterogeneous samples. Therefore, an optimal analytical strategy consists of examining each individual cell present in a complex population, would that be a solid or a liquid biopsy, in a parallel manner. We refer to this strategy here as parallel single cell analysis or PaSCAl [5]. For other diagnostic applications, only a few cells are available. For preimplantation genetic diagnostics (PGS), for instance, typically one or two blastomeres are isolated from the developing embryos for genetic diagnosis and screening [6]. Similarly, the analysis of fetal cells in maternal blood is conducted on a limited amount of cells [7], and in case of bacteremia, only a small number of bacteria must be detected and characterized from whole blood samples.

Microfluidics is a mature technology for the manipulation, processing, and high-sensitive analysis of minute amounts of sample [8], and, as such, it has opened new avenues in the field of single cell analysis [9, 10]. Specifically, microfluidic technology has enabled the development of a toolbox for experimentation at the single cell level [9]; for single cell manipulation and isolation; for cell characterization and/or analysis at the single cell level [11, 12]; for single cell engineering [13, 14]; for stimulation of individual cells, by taking advantage for instance of the unique flow properties at the microscale [15, 16]; or for single cell culture [17]. While different formats have been reported for the realization of microfluidic platforms for (parallel) single cell analysis, depending on the targeted analysis scheme and the used readout technique, these platforms share the same common features. One particular essential function common to all platforms is the ability to manipulate, trap, and isolate individual cells, and different approaches have reported to that end [18]: using integrated microstructures [13, 19] for mechanical trapping of cells; electrical trapping of cells using either electrophoresis [20] or dielectrophoresis [21]; through the encapsulation of individual cells in nL-droplets [22]; using chemical patterns [23]; or using optical tweezers [16] (*see* **Note 1** for detailed examples of alternative platforms for parallel single cell analysis).

Here, we describe one specific microfluidic platform for parallel single cell analysis relying on mechanical trapping of cells using lateral microstructures. This platform includes a series of trapping sites for the capture of a series of individual cells (Fig. 1), and each trapping site is directly coupled to a separate side-channel for the in situ analysis of the cell molecular content. While the herein-reported platform, which comprises different layers of structures, is fabricated from an elastomeric material (PDMS or polydimethylsiloxane), similar structures can easily be produced from silicon, glass, or another polymer material. Interestingly, this platform is highly versatile since it is compatible with a great variety of analysis schemes, it lends itself well to the implementation of multistep analytical processes, and it is easily scalable for the analysis of 100's to 1000's of cells. For instance, after trapping, isolated cells can be stained in situ and subsequently imaged using bright-field or fluorescence microscopy, as well as high-resolution imaging. Similarly, cells can be exposed to soluble stimuli and their response followed in real time using imaging techniques. Alternatively, the cell membrane can be ruptured, and the intracellular content recovered in the individual channels located behind each trapping site for molecular analysis. Finally, several of these steps can be combined

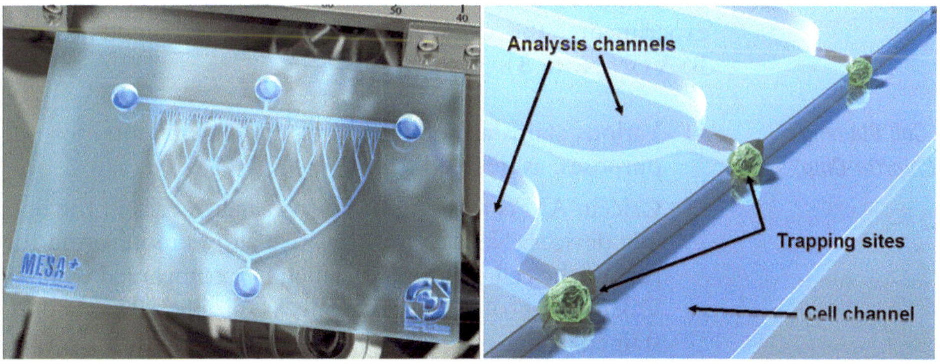

Fig. 1 Parallel single cell analysis—the concept. Artistic impressions of the microfluidic device comprising a main channel structure in which a cell suspension is introduced, and cells trapped in dedicated pockets present on one side of the channel. Each trapping site is connected to a separate side channel for eventual analysis of the cell molecular content. Artistic impressions by Nymus 3D

to access complementary information using different analysis strategies, or to identify changes at the molecular level after chemical stimulation of the cells. In this chapter, after a thorough discussion on parallel single cell trapping, different possible and independent steps of a complex analysis process are presented, and the utilization of various analysis protocols using both invasive and noninvasive approaches is demonstrated.

2 Materials

2.1 Cell Culture

1. Cells are grown in standard and commercially available culture medium supplemented with additives, as detailed below. The nature of the medium depends on the cell type.
2. P3X63Ag8 mouse myeloma cells are cultured in RPMI medium supplemented with 10 % v/v fetal bovine serum (FBS), 2 % v/v penicillin and streptomycin, 0.4 mg/mL fungizone, and 1 % v/v L-glutamine.
3. MCF-7 human breast cancer cells are cultured in DMEM medium supplemented with 10 % v/v fetal bovine serum (FBS), 2 % v/v penicillin and streptomycin, 0.4 mg/mL fungizone, and 1 % v/v L-glutamine.
4. A 1x trypsin solution in PBS buffer is employed to harvest MCF-7 cells for medium refreshment during culture or for cell sample preparation before experimentation.
5. Medium and serum are purchased from Invitrogen, and other supplements from Sigma-Aldrich.

2.2 Experimentation Buffer

All experiments are conducted in HEPES buffer (10 mM HEPES, pH = 7.4 composed of 10 mM HEPES, 140 mM NaCl, 2.68 mM KCl, 1.7 mM $MgCl_2$, 25 mM Glucose). All solutions to be inserted in the microfluidic device should be filtered (*see* **Note 2**). In some experiments, the experimentation buffer is supplemented with BSA (*see* **Note 3**).

2.3 Cell Staining (Off-Chip/On-Chip)

1. Various stains are employed in the experiments, for different purposes, as detailed below.
2. Calcein AM is first employed to visualize cells in the microfluidic devices. Next to this, this dye allows monitoring the rupture of the cell membrane, which is accompanied by leakage of Calcein out of the cells. Cells (typically 10^6 cells/mL) are stained using a Calcein AM working solution at 1 μg/mL (Invitrogen). This working solution is prepared by diluting 1000 times a Calcein AM stock solution (1 mg/mL in anhydrous DMSO) in culture medium.
3. Propidium Iodide (PI) is a cell-impermeable DNA-intercalating dye, which is widely used to detect damages in the cell membrane,

and therefore here cell membrane poration. PI is used at a concentration of 10 μg/mL. A PI working solution is prepared by diluting an aqueous stock solution (in DI water) in HEPES buffer (10 mM, pH 7.4).

4. The cell nucleus is stained in some experiments with Hoechst 33342 (Invitrogen). Hoechst is used at a concentration of 1 μg/mL for staining of a cell suspension at ca. 10^6 cells/mL, and a working solution is prepared by diluting 1000 times in culture medium a stock solution (1 mg/mL in anhydrous DMSO).
5. EpCAM is a receptor specifically expressed by cancer epithelial cells, such as MCF-7 cells, which can therefore be identified using a fluorescently labeled anti-EpCAM probe. A 1 μg/mL PE-anti-EpCAM solution in HEPES (10 mM, pH 7.4) is used to stain and identify cancer cells in a mixed MCF-7: P3X63Ag8 cell population.
6. In one series of experiments, the cell RNA content is stained with a selective probe, SYTO RNA Select. An intermediate SYTO RNA Select solution is prepared by diluting 1 μL of a stock solution (5 mM in anhydrous DMSO), in 1 mL of medium, which has been pre-warmed at 37 °C. Following this, the working staining solution is obtained by diluting the intermediate solution five or ten times with cell culture medium pre-warmed at 37 °C to yield a final SYTO RNA Select concentration of 0.5 or 1 μM.

2.4 Cell Fixation

One series of experiments is performed on fixed cells (*see* **Note 4**). Cell fixation is achieved here using an ice-cold methanol (−20 °C) solution (*see* **Note 5**), consisting of a 1:9 fresh PBS:methanol mixture, and which has been kept for at least 30 min at −20 °C.

2.5 Permeabilization/Lysis Experiments

1. Different approaches—relying on either physical or chemical means—are evaluated here for cell membrane permeabilization (*see* **Note 6**), for either permanent cell rupture or reversible membrane poration, as detailed in the following.
2. Cell chemical lysis is performed using a detergent solution based on 1 wt% LiDS (Sigma-Aldrich) in HEPES buffer (10 mM, pH 7.4).
3. Chemical (reversible) membrane poration is carried out using a 10 μg/mL digitonin solution in Ca^{2+}-free buffer (*see* **Note 7**), which is here HEPES buffer (10 mM, pH 7.4). The buffer is also supplemented with PI (10 μg/mL) for visualization of the cell membrane permeabilization.
4. Physical cell lysis relies here on the application of a high electric field across the cells. These experiments are conducted using a cell suspension in plain experimental buffer (HEPES 10 mM, pH 7.4)

2.6 Flow Visualization

A fluorescent solution of fluorescein sodium salt (Sigma-Aldrich) at a concentration of 1 mM in HEPES buffer (10 mM, pH 7.4) is utilized for visualization of the flows in the channels.

2.7 Microfluidic Device

1. All experiments are conducted using microfluidic devices fabricated from an elastomer material (PDMS or polydimethylsiloxane) which are bonded to a glass substrate.
2. All devices share the same main features (Fig. 2), which are: (1) a main channel, (2) a series of lateral cell trapping structures placed on one side of the main channel, and (3) a series of analysis or side-channels, each of them being connected to an individual cell trapping site.
3. In all devices, the main channel has a height of 30–50 μm and a width of 100 μm. This main channel is connected to one single outlet and to one or multiple inlets for the introduction of the cell suspension and chemicals.
4. Each device comprises 16 or 32 independent trapping structures and associated analysis channels. All trapping sites consist of a half-circular shaped pocket (radius of ca. 10 μm) having the same height as the main channel, and equipped with a shallow and narrow structure to prevent the cells from entering the analysis channel. All analysis channels are connected to one single reservoir in which a negative pressure is applied for cell trapping.
5. Two main designs (**Design 1** and **Design 2**) have been developed for different series of experiments and different purposes (Fig. 2); they share the same main features while exhibiting

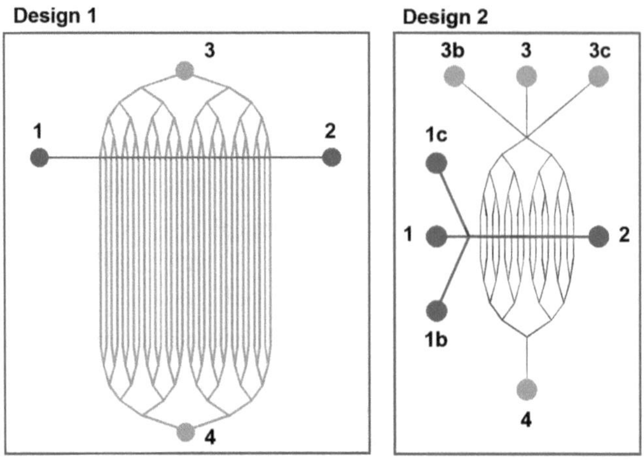

Fig. 2 Microfluidic device for parallel single cell analysis. Designs of the two devices utilized for the experiments presented here: Design 1 (*left*) and Design 2 (*right*) (see text for more explanations). Reservoirs are numbered, and these numbers are referred to in the text

small alterations regarding the design of the trapping sites and associated analysis channels, as discussed in the following.

6. **Design 1** (Fig. 2, left) uses a 2-layer fluidic structure [24].The height of the analysis channels is the same as that of the constrictions behind the traps, which is <5 μm (*see* **Note 8**). The analysis channels are made relatively wide (e.g., 100-μm width) to avoid high pressure drops in the analysis channels (*see* **Note 9**). In this design, the constricted section has a length of 30–40 μm and a width of 4–7 μm.
7. **Design 2** (Fig. 2, right) relies on a 3-layer fluidic structure [5]. The constriction behind the trap presents a height of 2 μm only to ensure cells cannot squeeze themselves into the analysis channel upon trapping. The analysis channel has a height of 10 μm and a width of 10 μm. In design 2, the length of the constricted section ranges from 0 to 30 μm.
8. All microdevices include, next to the array of analysis channels, a mirrored array of side-channels on the other side of the main channel, and these mirrored channels are also connected to one common reservoir. For cell electrical lysis, external electrodes are inserted in the common reservoirs of these two arrays of side-channels, and the electrical path between the electrodes for each cell is equivalent, which allows applying the same voltage on each cell.
9. Fig. 2 presents the two main designs used for the experiments presented in this chapter, and reservoirs have been numbered for both designs.

2.8 Experimental Setup

1. All experiments are conducted on an inverted microscope (Olympus IX51) equipped with an epi-fluorescence unit, suitable filter cubes, a camera (ColorView II, Olympus Soft Imaging Solutions), and connected to a computer for monitoring of the experiments and image acquisition. Dedicated software (AnalySIS docu 5.0) is used for image acquisition.
2. Fluids are actuated in the microfluidic system using different pieces of equipment, as detailed in the following.
3. Cell trapping in the microfluidic device is monitored using a pressure-controlled system (Maesflo, Fluigent) operated in the negative pressure range and controlled via dedicated software (Maesflo 4c v0.6.1).
4. Chemicals are introduced in the microfluidic device at a given flow-rate in the main channel, using a syringe-pump (Harvard PHD 22/2000, Hugo Sachs Elektronik-Harvard Apparatus GmbH), which is used in the suction mode.
5. A HV source (IBI 411, IBIS Technologies BV) is employed for cell electrical permeabilization and establishment of an

electro-osmotic flow in the analysis channel for cell content retrieval. The HV is applied using Pt electrodes, which are inserted in the reservoirs 3 and 4 of the microfluidic device.

3 Methods

3.1 Device Fabrication

The microfluidic devices are fabricated using soft-lithography techniques from an elastomer material (PDMS or polydimethylsiloxane). First, a silicon/SU-8 mold is fabricated in the cleanrooms of the MESA+ Institute for Nanotechnology and treated with a hydrophobic FDTS (1H,1H,2H,2H-perfluorodecyltrichlorosilane) coating to facilitate the release of the PDMS structures. Next, a 10:1 PDMS pre-polymer:curing agent mixture is prepared, thoroughly degassed and poured on the mold. PDMS is cured overnight in an oven at 60 °C. After curing, the PDMS structures are released from the mold, and reservoirs are punched with a sharp blunt needle. The PDMS layer is subsequently cleaned with isopropanol, carefully dried and exposed to oxygen plasma for its activation before bonding to a clean glass substrate. Figure 3 presents the picture of a microfluidic device (Design 2), as well as an enlarged view of one trapping structure (Design 2).

3.2 Device Preparation

Before experimentation, the microfluidic channels are incubated with a filtered BSA solution (5 % in HEPES buffer, 10 mM, pH 7.4) for 2 h to prevent undesired cell adsorption on the channel walls, followed by thorough rinsing with HEPES buffer.

3.3 Cell Culture and Cell Sample Preparation

1. Cells are cultured in their respective medium (RPMI for P3X63Ag8 and DMEM for MCF-7 cells), and the medium is typically refreshed twice a week. Specifically, for P3X63Ag8 cells, the cell suspension is collected from a standard T25 culture flask, centrifuged to remove dead cells and debris and ca.

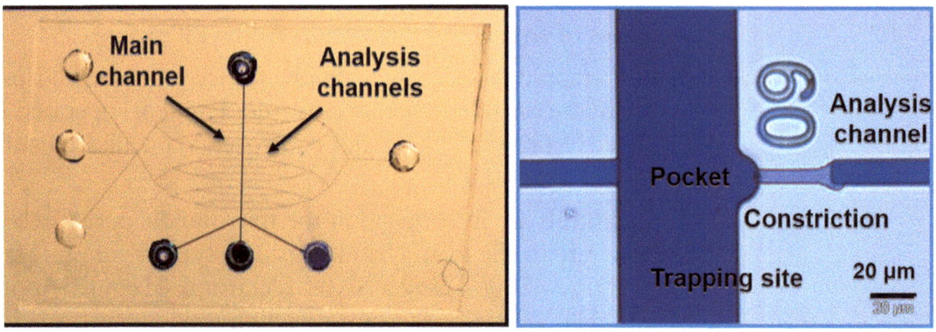

Fig. 3 Microfluidic device for parallel single cell analysis. Pictures of an actual device (Design 2) (*left*) whose microfluidic structures have been filled with *blue ink* for visualization purpose and of one trapping site (*right*), showing the three-layer structure

one sixth of the cells is resuspended in 5 mL of fresh medium. MCF-7 cells, which are adherent cells, are first washed with PBS, and detached from the culture flask surface using 1 mL of 1x trypsin for 2 min at 37 °C. Following this, the resulting cell suspension in PBS is treated in the same way as the P3X63Ag8 cells.

2. Depending on the experiments and the specific targeted goal, cells are exposed to different preparation steps, as discussed in the following for three specific experiments.
3. First, experiments are conducted both on live cells and fixed cells (*see* **Note 4**), fixation being performed prior to the introduction of the cells in the microfluidic device. To that end, cells are slowly and carefully resuspended in an ice-cold 1:9 PBS:methanol solution while gently pipetting to avoid cell aggregation. After ca. 10 min incubation at 4 °C, the cells are washed with cold PBS at least twice, and resuspended in experimentation buffer.
4. For cell trapping experiments, cells are stained with Calcein and/or Hoechst, these dyes being used to easily visualize the cells in the channels and trapping structures. In one series of experiments, these two dyes allow distinguishing two cell populations. Specifically, when a mixed MCF-7:P3X63Ag8 cell sample is used, one cell type is stained with one dye (Hoechst only, here) while the other cell type is stained with both probes (Hoechst & Calcein). For both dyes, the staining procedure is the same: the cell sample is resuspended in the staining solution, and incubated for 30 min at 37 °C. Thereafter, the cells are washed at least twice in HEPES buffer (10 mM, pH 7.4) to remove the excess of dye.
5. For cell lysis and cell content recovery experiments, cells are only loaded with Calcein AM using the same protocol as described above. Calcein is utilized here to probe pore formation in the membrane, which gives rise to Calcein leakage out of the cells, and allows monitoring cell content retrieval in the analysis channels.
6. One specific series of experiments aims at recovering the nucleic acid content of the cells. For these experiments, cells are exposed to a dual staining procedure using Hoechst 3342 and SYTO RNA Select, to separately label their DNA and RNA contents, respectively. Typically, 1 mL of cell solution is resuspended in the staining solution, which consists of 1 mL of warm medium containing the SYTO RNA Select probe and to which 1 μL of Hoechst stock solution is added. After 30–45 min incubation, the cell sample is washed 1–2× with buffer. This dual staining protocol can be applied to both living and fixed cells.

3.4 Cell Loading

1. Cells are loaded in the device using the passive pumping technique [25] (*see* **Note 10**). Specifically, a 15 μL droplet of HEPES buffer is introduced in reservoir 2, and a 1.5 μL droplet of a cell suspension in HEPES buffer in reservoir 1. Thereby, a flow is created from reservoir 1 to reservoir 2, and cells are transported with the flow in the main channel with a mild velocity of 100–200 μm/s.
2. Alternatively, the cell suspension can be deposited in one inlet reservoir, and suction applied using a syringe pump connected to reservoir 2 to create a constant and well-defined flow in the main channel.

3.5 Cell Trapping

While loading the cell suspension (10^6 cells/mL) in the microfluidic device and the main channel, a mild negative pressure (−30 mBar) is applied from the common back reservoir (#4) to attract cells in the different trapping sites, using a Maesflo pressure controller. The applied pressure is controlled via dedicated software with a great precision (as low as 0.345 mBar) and can easily be fine-tuned, depending on the exact design of the microfluidic structures. Once a cell is captured in a trap, the trap is blocked provided there is a good sealing between the cell and the trap, so that no other cell can be attracted in the trapping site. Once all traps have been filled with cells (ideally with one single cell), the pressure is reduced and maintained to −10 mBar to keep the cells in the traps, while flushing other solutions in the main channel.

Typically, using these conditions, parallel cell trapping is achieved within 2–5 min in a device equipped with 16–32 traps, and both proposed designs (**Design 1** and **Design 2**) yield similar trapping results. Figure 4 shows an array of individual cells trapped in a microfluidic device (Design 2). The trapping efficiency is characterized here with respect to the number of cells isolated per trapping site (0, 1 or more cells) (*see* **Note 11**). For both designs, the single cell trapping yield is high, and typically amounts to 70 % (as tested for several devices and different experimenters).

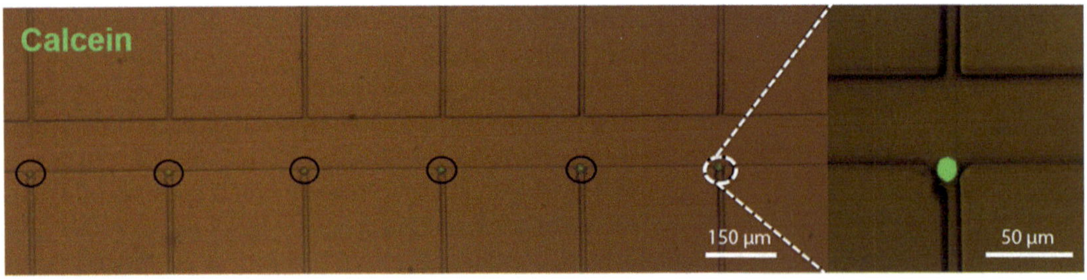

Fig. 4 Parallel single cell trapping. Picture of a portion of the main channel (Design 2) showing the trapping of a series of seven individual cells in separate trapping sites (*left*) and enlarged view of one captured cell. Cells are stained with Calcein for visualization purposes

Interestingly, the exact design of the trapping structure is found to have an effect on the single cell trapping yield: the shortest constriction structure gives the best single cell trapping yield (85 %), and other constrictions lengths result in similar trapping yields. Next, in typically 10–20 % of the trapping sites, multiple cells are found [5].

For in vitro diagnostic applications, or more in general, for medical analysis, the cell sample to be analyzed is in nature heterogeneous, and comprises cells of different sizes. To simulate a heterogeneous sample consisting of cells with different sizes, and to demonstrate the ability to work with mixed cell populations, a 1:1 MCF-7:P3X63Ag8 cell sample is prepared off-chip. Prior to this, MCF-7 cells are stained with both Hoechst and Calcein, while P3X63Ag8 cells are stained with Hoechst only before mixing the cells in a 1:1 ratio. The cell trapping efficiency and single cell trapping yield is not altered when working with mixed samples comprising two types of cells with slightly different sizes. Furthermore, this experiment confirms the suitability of the proposed device to handle mixed cell populations, which is particularly attractive in the future for the analysis of CTC samples that always contain residual WBCs.

3.6 On-Chip Cell Staining

After trapping in the herein presented microfluidic device, cells can be exposed to various soluble factors in a highly controlled manner by taking advantage of the main microfluidic channel, for their chemical stimulation or their staining. The latter approach allows detecting or identifying targeted cells in a mixed cell population, using specific (fluorescent) staining agents. For instance, cancer cells can be identified in a heterogeneous population using EpCAM (Epithelial Cell Adhesion Molecule), a marker expressed by epithelial cancer cells only, and this approach has proven to be powerful to identify CTCs after their purification from blood, or to detect CTCs in complex samples [26].

Here, to demonstrate the possibility to identify specific cells in a mixed population in our microfluidic device, a heterogeneous cell sample combining EpCAM-positive MCF-7 cancer cells and EpCAM-negative P3X63Ag8 cells in a 1:1 ratio is prepared and trapped as described above. Following this, and while maintaining a mild negative pressure (−10 mBar) from the back reservoir (#4), a staining solution containing Hoechst and PE-Anti-EpCAM is injected in the main channel for on-chip cell staining. After 30–60 min incubation with the staining solution followed by on-chip cell washing with HEPES buffer, cells are imaged in the device using fluorescence microscopy. As shown in Fig. 5, after on-chip staining, MCF-7 cells exhibit both blue and red labels, for Hoechst and PE, respectively, while P3X63Ag8 are only labeled with Hoechst (blue label).

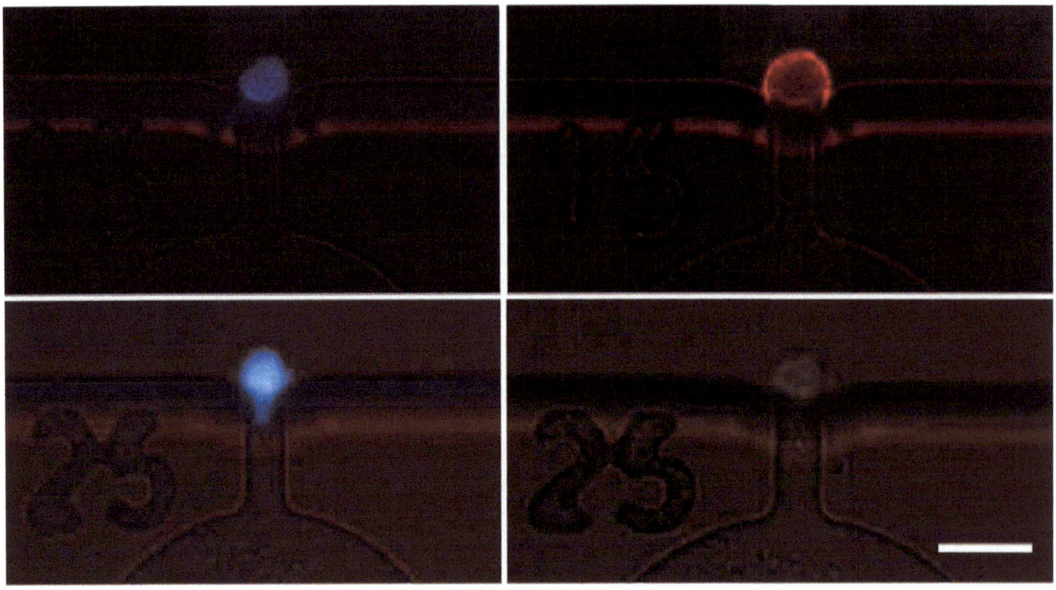

Fig. 5 In situ cell staining in the microfluidic device to identify cells in a mixed cell population. Enlarged view of two cells (P3x63Ag8 (*Bottom*) and MCF-7 (*Top*)) Differential staining of P3x63Ag8 and MCF-7 cells in a device, using Hoechst (*Left*) that label all cells, and an PE-conjugated antibody aiming at the membrane marker EpCAM (*right*) which is solely expressed by epithelial cancer cells (MCF-7 cells, here) [5]. Figure reprinted from [5] John Wiley and Sons with authorization

3.7 Cell Permeabilization/Lysis

1. In a second possible and more invasive analytical scheme, the cell membrane is ruptured to access the intracellular content. Different approaches are conceivable for cell membrane permeabilization, e.g., for reversible/temporary pore formation and sampling from the cell or irreversible cell poration, and using either chemical or physical means (*see* **Note 6**). Importantly, all these different approaches are compatible with the herein-proposed microfluidic platform, as detailed in the following.

2. For reversible permeabilization of live cells, the non-ionic detergent digitonin is utilized (*see* **Note 7**). Digitonin forms a complex with cholesterol present in the plasma membrane, and upon aggregation of several digitonin-cholesterol complexes, a membrane-spanning pore is formed [27] with a diameter around 8–10 nm. Here, the digitonin solution is applied in reservoir 1 (inlet), and pumped at a constant flow-rate using a syringe-pump connected to reservoir 2, while maintaining a mild negative pressure (−10 mBar) from the common back reservoir (#4) using the Maesflo system to prevent cells from escaping the trapping structures. The digitonin solution is supplemented with the cell-impermeable fluorescent probe, PI, to study the process of cell membrane poration. Upon direct exposure to digitonin, the cell opens up, becomes positive to PI while slowly swelling, which reflects the differences in osmolarity

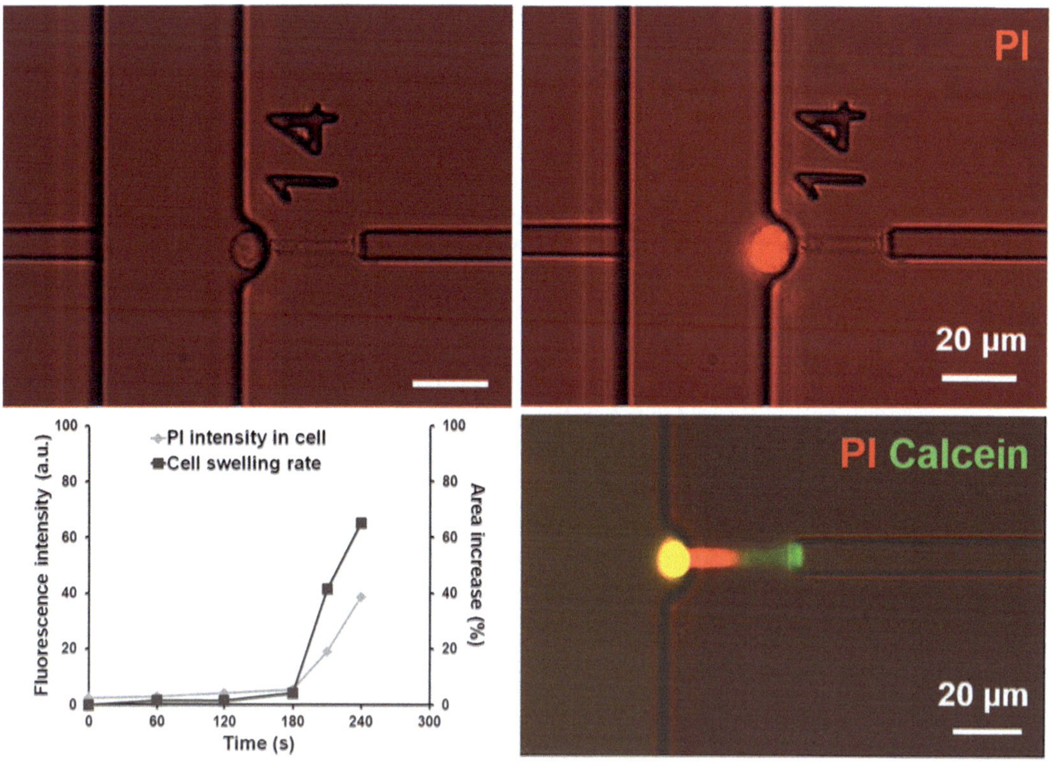

Fig. 6 Chemical (reversible) poration using digitonin of P3x63Ag8 cells in a microfluidic device (Design 2). Pictures of an individual cell in a trapping site during the permeabilization process using 10 μg/mL of digitonin in HEPES buffer (10 mM, pH 7.4). The buffer is supplemented with PI to monitor the process of cell membrane poration. The pictures taken at 0 s (*top left*) and 240 s (*top right*) indicate swelling of the cell and entry of PI in the cell after exposure to digitonin, which is also quantified over time for this particular cell (*bottom left*). Image of a cell previously loaded with Calcein, which is tightly secured in a trapping site, and simultaneously exposed to digitonin and PI as before. Interestingly, no release of Calcein is detected and PI slowly diffuses into the cell from its main channel side. Figure adapted from John Wiley and Sons with authorization [5]

in the buffer and intracellular medium (Fig. 6). Furthermore, after typically 3 min exposure, cell swelling and PI uptake dramatically increases (Fig. 6, bottom left) PI being still mostly found on the side of the cell facing the main channel, where digitonin is introduced. Surprisingly, no significant leakage of Calcein, which was loaded into the cells prior to experiments, is observed.

3. In a second series of experiments, cells are chemically lyzed via their exposure to a detergent solution, which induces dissolution of their cytoplasmic membrane. As a detergent, LiDS is used here: LiDS is milder than SDS, so that the process of cell lysis is slower, which allows progressive recovery of the cell content in the side channels. Furthermore, LiDS has been reported to be selective to the cell cytoplasmic membrane and to not induce lysis of the intracellular membranes, which is

particularly interesting for the separate isolation and analysis of the cell cytoplasm and its organelles. The LiDS solution (concentration of 1 wt%) is flushed in the main channel while maintaining a negative pressure from the back reservoir, not only to keep the cells in the trap, but also to possibly isolate the cell content in the analysis channels. In this experiment, P3X63Ag8 cells, which have previously been fixed and stained with Hoechst and RNA SYTO Select, are used. First, as in the presence of digitonin, the cell is swelling upon exposure to LiDS. Furthermore, and more interestingly, the nucleus seems to remain intact after exposure to LiDS 1 wt%, while the cytoplasm content is found back in the analysis channel, which is evidenced, respectively, by the remaining blue stain in the cells, and the presence of green dye in the channel in the vicinity of the trapping site (Fig. 7).

4. In a last experiment, the cell membrane permeabilization is demonstrated using an electric field (*see* **Note 12**) and the so-called technique of electroporation [28]. External electrodes are inserted in reservoirs 3 and 4, for the application of a HV across the cells captured in the trapping sites for their permeabilization, as well as in the analysis channels for cell content extraction (*see* **Note 13**). Typically, voltages of 200–300 V are applied (*see* **Note 14**). Such voltages give rise to an electro-osmotic flow velocity in the side channels of 10's μm/s (measured velocity of 78 μm/s upon application of 300 V), and to a theoretical voltage drop across the trapping sites higher than 1 kV/cm, as required for the creation of pores in the cell membrane. Within less than one second, cell membrane poration is observed, as evidenced by the release of Calcein, which has previously been loaded in the cells. More importantly, under these conditions, the intracellular content is selectively extracted in the analysis channels (Fig. 8).

3.8 Cell Content Retrieval

1. Two approaches have been explored to retrieve the cell content in the analysis channels during lysis for further cell content analysis (*see* **Note 15**).
2. In a first approach, demonstrated here in combination with the chemical lysis of cells using LiDS (see Subheading 3.7 **step 3** above and Fig. 7), the cell content is selectively extracted in the analysis channels by applying a mild negative pressure from the common suction port (#4) while exposing the cell to the detergents.
3. In a second approach where lysis occurs upon application of an electric field between reservoirs 3 and 4 and across the cell (see Subheading 3.7 **step 4** above), the same electric field allows creating an electro-osmotic flow in the side-channels, which is used here for the controlled transport of the cell content in the analysis channels (*see* **Note 14e** and Fig. 8).

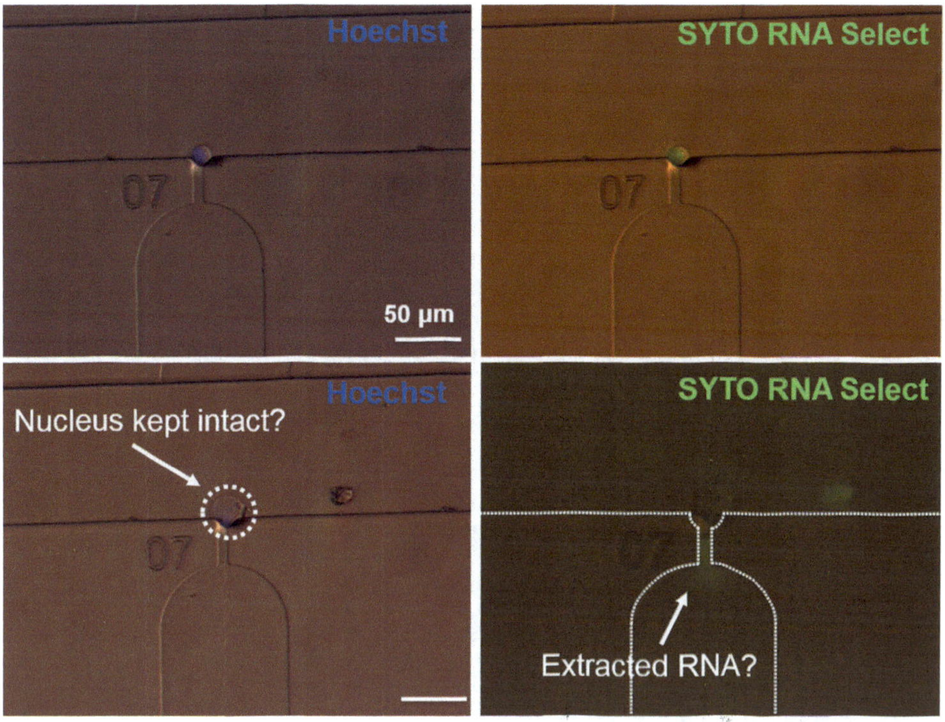

Fig. 7 Chemical permeabilization sing LiDS of P3x63Ag8 cells in a microfluidic device (Design 1) combined with pressure-based extraction of the cell content in the analysis channels. Pictures of an individual cell in a trapping site before (*top*) and after (*bottom*) the permeabilization process using 1 wt% LiDS, extracted from a time-lapse movie. Cells were stained with both Hoechst (*left*) and RNA SYTO Select (*right*) prior to their injection in the microfluidic device and their trapping, to monitor both processes of cell lysis and cell content recovery. The pictures after lysis (*bottom*) show swelling of the cells upon exposure to LiDS for ca. 60 s, together with extraction of the RNA in the analysis channel, and seem to indicate that the nucleus has been kept intact

4 Notes

1. Other platforms and approaches have been reported for (parallel) single cell analysis that rely on different strategies for single cell trapping. A first platform aiming at parallel single cell analysis consisted of an array of microfabricated wells [19, 46] in which individual cells could be trapped, provided optimization of the structure sizes and of the seeding protocol. In an alternative approach, a 2D array of lateral trapping structures was placed in a microfluidic chamber [29, 47]; this array was demonstrated for live cell imaging and monitoring of specific biological processes.
2. All solutions to be injected in the microfluidic devices must be carefully filtered beforehand to remove all debris that could easily clog the small features as found here in the trapping structures. In our experiments, we did observe clogging of

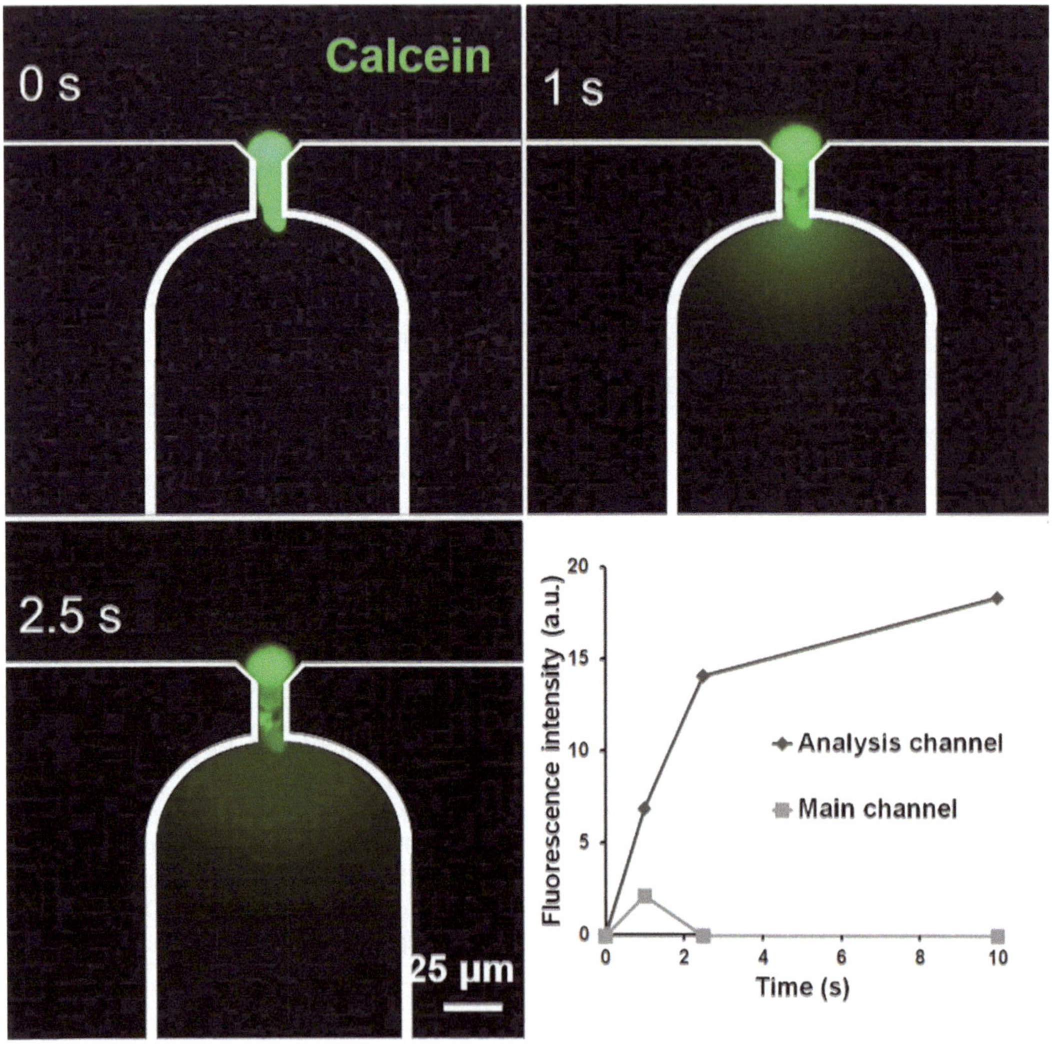

Fig. 8 Electrical permeabilization of P3x63Ag8 cells in a microfluidic device (Design 1) combined with EOF-based extraction of the cell content in the analysis channels. Pictures extracted from a time-lapse movie, illustrating the cell membrane poration and simultaneous recovery of the cell content in the analysis channel (Top *right* and *left*, and *bottom left*). Cells were stained with Calcein prior to their injection and trapping in the device to monitor both processes of membrane poration and content extraction. Variations in the green fluorescence intensity as a function of time in the analysis channel and main channel for this particular cell, on which a HV of 300 V has been applied (*Bottom right*). Figure reprinted from [5] John Wiley and Sons with authorization

some traps, even after careful filtration, which contributed to a decrease in the trapping yield.

3. If no special care is taken, cells in suspension tend to aggregate to form clusters. Cell aggregation results in trapping of multiple cells and not of individual cells, as targeted. To avoid this, cells in the solution must be resuspended through careful pipetting. Alternatively, the cell experimentation buffer can be

supplemented with a few % of BSA or accutase to prevent cell aggregation.

4. For most applications, live cells are preferred, especially for imaging-based monitoring of cellular processes [29]. However, fixed cells can be more advantageous than live cells: they are easier to trap since their membrane has been made more rigid.
5. For cell fixation, ice-cold methanol is preferred here since it does not affect intracellular biomolecules, as formaldehyde would do.
6. Two lysis schemes have been tested and validated here, using detergents (chemical lysis) and an electric field (physical lysis). Other schemes are also possible, such as the use of a laser pulse which induces the formation of a cavitation bubble [32, 33], surface acoustic waves [34], or mechanical lysis using sharp nanostructures [35] or beads. In general, physical means (e.g., electric field, sonolysis, etc.) are preferred since they can be made cell-specific. Lysis also occurs on a faster time scale than when using chemicals, which is particularly interesting to study fast processes in cells such as signaling pathways [36]. Furthermore, loss of biological materials through diffusion is limited. Finally, chemicals represent a source of contamination, and their use results in an overall dilution of the cell content sample.
7. The process of pore formation using digitonin is sensitive to the presence of calcium ions: pore formation induced by digitonin only occurs when no Ca^{2+} is present in solution [31], so that the buffer should not comprise any calcium ions. Interestingly, digitonin-induced pore formation is reversible and the addition of a calcium salt in solution assists resealing of the pore [31], a capability that has not been explored here, but would allow sampling out of cells in a seamless manner.
8. Different types of trapping structures have been tested here, with several alterations in their design (e.g., number of layers of structures, exact dimensions of the constricted section, etc.). In general, a compromise must be found between the ease of fabrication and cell trapping efficiency: larger structures are easier to produce, but they let more cells go through. In contrast, small structures are advantageous as discussed below (*see* **Notes 16** and **17**) since they offer better sealing of the cells in the trapping sites. Working with fixed cells allows the use of capture sites with larger dimensions, while experimentation using live cells calls out for smaller dimensions.
9. All devices reported here have been fabricated using PDMS. This elastomeric material is particularly attractive for prototyping, and the production of single-use devices, especially in an academic environment. However, PDMS suffers from a number of limitations. For instance, in Design 1 (Fig. 2, left), the analysis channels have a high aspect ratio (100-μm width and

2-μm height), and their dimensions were chosen to avoid cell squeezing in the analysis channels while limiting the pressure drop in these channels. However, when no particular care was taken, channel collapse was observed during bonding of the devices, and the analysis channels were "lost." Similarly, the realization of small structures can be challenging due to the thermal sensitivity of the PDMS material. This thermal sensitivity becomes a significant issue, if the PDMS fluidic layer must be aligned to other structures on the glass substrates. For instance, for cell electrolysis, integrated electrodes would be preferred (*see* **Note 13**), which is not conceivable when using PDMS. Last but not least, extraction of the cell content in the analysis channels relies in one series of experiments on the establishment of an electro-osmotic flow (EOF). However, the surface charges on the PDMS walls are ill-defined, so that the resulting EOF flow was not well controlled and subsequently not reproducible.

10. Cells have been introduced in the system using two main approaches, both involving the deposition of the cell suspension in the main inlet reservoir of the device (#1) and creation of a flow from the main outlet reservoir (#2): either using the passive pumping technique [25] or using a syringe-pump that allows the establishment of a steady flow in the main channel. Alternatively, cells could be inserted using a pumping system connected to the inlet reservoir, such as a syringe-pump or a pressure-controlled system (e.g., Maesflo system operated in the positive pressure mode). However, the latter approaches suffer from one major issue: cells tend to sediment in syringes or vials, so that active shaking is required to limit this. Therefore, here, we preferred using strategies involving aspiration of the cells from the inlet reservoir (#1).
11. Here, the trapping yield was defined with respect to the number of trapping sites present in the device and using the number of traps in which one or multiple cells were captured. Typically, for each experiment, 100's to 1000's of cells were injected in the device. However, for in vitro diagnostic applications, when a reduced number of cells is available, e.g., a handful of cells for CTC analysis, all cells injected in the device must be captured. Therefore, the trapping yield should be corrected for the number of cells injected in the device, and the trapping protocol must be altered to ensure every single cell inserted in the device is captured.
12. Electrical lysis has been demonstrated here. Compared to other physical lysis techniques, the process can be made reversible or irreversible by tuning the electric field parameters (amplitude of the signal, or time duration of the treatment). Furthermore, while applying an electric field for rupturing the cell membrane, an electro-osmotic flow is created in the side channels, which means that the cell membrane permeabilization

can directly be combined with the recovery of the cell content for further analysis.

13. Here, to create an electric field, external electrodes are inserted in the reservoirs 3 and 4 of the device and all trapping sites are exposed simultaneously to the same electric field. Using integrated and individually addressable electrodes in the future would allow making the process cell-specific, while using much lower voltage to reach the kV/cm threshold required for cell membrane poration [13]. It should be however noted that integrating the electrodes in the device would most probably imply changing materials to fabricate the devices, since the alignment of electrode structures with a series of micrometer-sized trapping structures using PDMS would be highly challenging or even not possible (*see* **Note 9**).
14. A DC voltage a few 100's V has been applied in these experiments for both cell electrical lysis and creation of an electro-osmotic flow. Under these conditions, the cells were porated but not entirely lyzed. A better strategy would rely on the use of a series of short pulses of a higher voltage for lyzing the cells, followed by a much lower DC signal for the EOF.
15. While experiments presented here are limited to cell trapping followed by cell staining and exploration of different permeabilization approaches, the herein-proposed platform is ideally suited for a great variety of analytical schemes. Cell imaging was demonstrated here to recognize cancer cells in a mixed population, but it can easily be extended to in situ FISH analysis, for instance, to determine gene copy numbers in individual cancer cells [37] or to examine possible aneuploidy or detect chromosome imbalance in blastomeres [38]. Imaging could also be performed using label-free and less invasive techniques such as Raman spectroscopy [39], provided the bottom substrate is replaced for CaF_2. After trapping, cells could alternatively be exposed in a controlled way to various drugs or chemical stimuli, and their response recorded in situ in the device, e.g., using time-lapse imaging, or after cell lysis to study specific changes in their intracellular content [40]. After cell permeabilization and retrieval of the content of individual cells in separate side channel, molecular analysis can be performed, e.g., to examine gene expression profiles using RT-PCR [41] and/or integrated mRNA microarrays, to unravel mutation patterns in the genome of individual cells using whole genome amplification [42], to examine the activity or expression level of specific proteins using enzymatic assays [43], capillary electrophoresis (CE) [44] or even integrated antibody arrays [45].
16. In these experiments, two trapping modes were observed, while using the same conditions for trapping and the same trapping structures. Specifically, cells could be either sitting in

front of the traps, or partially sucked in the traps. While no specific parameters could be identified that would influence the way cells would be positioned in the capture site, this trapping mode has an impact on the overall analytical process: for instance, on the risks to capture multiple cells in one site, or to release cells when applying a flow in the main channel; on the efficient recovery of the cell content in the analysis channels without any risk of loss or contamination; or on the efficiency of the electrical lysis.

17. Cells are trapped here by applying a negative pressure from the common back reservoir [4] to attract cells in the trapping sites. While this approach is preferable to have good sealing of the cells in the trapping structure that is important for multiple reasons (*see* **Note 16**, above), it also requires the use of smaller structures to limit cell loss through their passage as a whole in the analysis channels (*see* **Note 8**). Specifically, our experiments revealed that the constricted section should be no higher than 2–3 μm when working with live cells. In another approach, inspired by the work of Seki's group [30], a focusing flow could be used in the main channel to assist cell capture by pushing individual cells in the trapping sites. This milder approach is preferred to maximize the trapping yield, with respect to the number of cells available for the analysis and injected in the device, and to limit the stress applied on the cells. However, using this approach, there exists a risk that cells are not tightly trapped, which raises other issues, as discussed above (*see* **Notes 8** and **16**).

While these bottom and lateral trapping approaches are easily scalable for the capture of thousands of cells, they are exclusively limited to imaging of cells. These two platforms have therefore been upgraded to become compatible with intracellular molecular analysis, through the confinement of each individual cell in a closed compartment. For the former platform, a glass microscope was placed on the microwell array, after cell trapping to create a series of closed microreactors. This strategy has, for instance, been reported by Gong et al. for targeted gene expression analysis in 125 pL microchambers [48] using RT-PCR. Compartmentalization can also be achieved by using microchambers and/or valves realized by soft lithography. Individual cells have been brought into nL-chambers with the help of an optical tweezer [49], and lyzed in situ. In this device, the capture chamber was decorated with an antibody array for analysis of the number of protein copy number (here, for p53) in the lysate of the individual cells. Entirely closed chambers with a well-defined volume can easily be created using PDMS valves, which allow enclosing individual cells. Marcus et al. have, for instance, pioneered this strategy for mRNA analysis from individual cells [41].

More recently, Yang et al. reported whole genome amplification from individual cells in such microchambers for eventual genome sequencing and mapping of mutation patterns, at the single cell level [42]. In an alternative approach, the actuation layer comprised a more complex doughnut-like structure, which was pushed down to yield a closed or partially closed chamber [50]. Cells trapped in this chamber were lyzed chemically and their NADPH content analyzed in situ.

Individual cells can also easily be isolated in (sub)-nL droplets produced in a continuous immiscible oil phase [22] in a very high-throughput manner (up to 1000 droplets per second). In a seminal paper, Brouzes et al. demonstrated the potential of this droplet microfluidic platform for screening of a library of drugs on individual cells [51]. More recently, the same strategy has been applied to monitor drug uptake and cell viability in cells either sensitive or resistant to a specific treatment [52].

Finally, in a radically different approach, high-throughput single cell analysis was achieved in a flow-through manner, using optical lysis of individual cells at a channel intersection followed by online CE separation of their content and detection of targeted compounds using fluorescence microscopy [53], to eventually elucidate alterations in kinase activation and intracellular signalization pathways [36].

Acknowledgment

The work presented in this chapter has been conducted in part by Floris van den Brink and Elmar Gool. Johan Bomer, Daniël Wijnperlé, and Hans de Boer are thanked for technical assistance (mold fabrication and development of the setup). Prof. Edwin Carlen is acknowledged for his supervision in the first year of the project. This work was supported by NanoNext NL, a micro and nanotechnology consortium of the Government of the Netherlands and 130 partners as well as by the company OGT (Oxford Gene Technology).

References

1. Haselgrubler T, Haider M, Ji BZ et al (2014) High-throughput, multiparameter analysis of single cells. Anal Bioanal Chem 406(14):3279–3296
2. Young EWK (2013) Cells, tissues, and organs on chips: challenges and opportunities for the cancer tumor microenvironment. Integr Biol-Uk 5(9):1096–1109
3. Tian TH, Olson S, Whitacre JM, Harding A (2011) The origins of cancer robustness and evolvability. Integr Biol-Uk 3(1):17–30
4. Ligthart ST, Bidard FC, Decraene C et al (2013) Unbiased quantitative assessment of Her-2 expression of circulating tumor cells in patients with metastatic and non-metastatic breast cancer. Ann Oncol 24(5):1231–1238

5. van den Brink FT, Gool E, Frimat JP et al (2011) Parallel single-cell analysis microfluidic platform. Electrophoresis 32(22):3094–3100
6. Harper JC, SenGupta SB (2012) Preimplantation genetic diagnosis: state of the ART 2011. Obstet Gynecol Surv 67(6):347–348
7. Steele CD, Wapner RJ, Smith JB et al (1996) Prenatal diagnosis using fetal cells isolated from maternal peripheral blood: a review. Clin Obstet Gynecol 39(4):801–813
8. Whitesides GM (2006) The origins and the future of microfluidics. Nature 442(7101): 368–373
9. Le Gac S, van den Berg A (2010) Single cells as experimentation units in lab-on-a-chip devices. Trends Biotechnol 28(2):55–62
10. Sims CE, Allbritton NL (2007) Analysis of single mammalian cells on-chip. Lab Chip 7(4):423–440
11. Liberale C, Cojoc G, Bragheri F et al (2013) Integrated microfluidic device for single-cell trapping and spectroscopy. Sci Rep 3:1258. doi:10.1038/srep01258
12. Hochstetter A, Stellamanns E, Deshpande S et al (2015) Microfluidics-based single cell analysis reveals drug-dependent motility changes in trypanosomes. Lab Chip 15(8):1961–1968
13. Valero A, Post JN, van Nieuwkasteele JW et al (2008) Gene transfer and protein dynamics in stem cells using single cell electroporation in a microfluidic device. Lab Chip 8(1):62–67
14. Adamo A, Jensen KF (2008) Microfluidic based single cell microinjection. Lab Chip 8(8):1258–1261
15. Takayama S, McDonald JC, Ostuni E et al (1999) Patterning cells and their environments using multiple laminar fluid flows in capillary networks. P Natl Acad Sci USA 96(10):5545–5548
16. Eriksson E, Enger J, Nordlander B et al (2007) A microfluidic system in combination with optical tweezers for analyzing rapid and reversible cytological alterations in single cells upon environmental changes. Lab Chip 7(1):71–76
17. Kobel S, Valero A, Latt J et al (2010) Optimization of microfluidic single cell trapping for long-term on-chip culture. Lab Chip 10(7):857–863
18. Le Gac S, Van den Berg A (2010) Cell capture and lysis on a chip. In: Bontoux N, Dauphinot L, Potier M-C (eds) Unravelling single cell genomics. vol RSC nanosciences & nanotechnology, RSC Publishing, pp 150–184.
19. Rettig JR, Folch A (2005) Large-scale single-cell trapping and imaging using microwell arrays. Anal Chem 77(17):5628–5634
20. Toriello NM, Douglas ES, Mathies RA (2005) Microfluidic device for electric field-driven single-cell capture and activation. Anal Chem 77(21):6935–6941
21. Taff BM, Voldman J (2005) A scalable addressable positive-dielectrophoretic cell-sorting array. Anal Chem 77(24):7976–7983
22. Clausell-Tormos J, Lieber D, Baret JC et al (2008) Droplet-based microfluidic platforms for the encapsulation and screening of Mammalian cells and multicellular organisms. Chem Biol 15(5):427–437
23. Thery M, Racine V, Pepin A et al (2005) The extracellular matrix guides the orientation of the cell division axis. Nat Cell Biol 7(10):947–U929
24. Le Gac S, de Boer H, Wijnperlé D et al Parallel single cell analysis on an integrated microfluidic platform for cell trapping, lysis and analysis. In: MicroTAS, the 13th international conference on miniaturized systems for chemistry and life sciences, Jeju Island, Korea, 2009. Royal Society of Chemistry.
25. Walker G, Beebe DJ (2002) A passive pumping method for microfluidic devices. Lab Chip 2(3):131–134
26. Rao CG, Chianese D, Doyle GV et al (2005) Expression of epithelial cell adhesion molecule in carcinoma cells present in blood and primary and metastatic tumors. Int J Oncol 27(1):49–57
27. Olofsson J, Bridle H, Jesorka A et al (2009) Direct access and control of the intracellular solution environment in single cells. Anal Chem 81(5):1810–1818
28. Neumann E, Schaeferridder M, Wang Y, Hofschneider PH (1982) Gene-transfer into Mouse Lyoma cells by electroporation in high electric-fields. Embo J 1(7):841–845
29. Di Carlo D, Aghdam N, Lee LP (2006) Single-cell enzyme concentrations, kinetics, and inhibition analysis using high-density hydrodynamic cell isolation arrays. Anal Chem 78(14):4925–4930
30. Yamada M, Kano K, Tsuda Y et al (2007) Microfluidic devices for size-dependent separation of liver cells. Biomed Microdevices 9(5):637–645
31. Miyamoto K, Yamashita T, Tsukiyama T et al (2008) Reversible membrane permeabilization of mammalian cells treated with digitonin and its use for inducing nuclear reprogramming by Xenopus egg extracts. Cloning Stem Cells 10(4):535–542
32. Gac SL, Zwaan E, van den Berg A, Ohl CD (2007) Sonoporation of suspension cells with

a single cavitation bubble in a microfluidic confinement. Lab Chip 7(12):1666–1672
33. Quinto-Su PA, Lai HH, Yoon HH et al (2008) Examination of laser microbeam cell lysis in a PDMS microfluidic channel using time-resolved imaging. Lab Chip 8(3):408–414
34. Salehi-Reyhani A, Gesellchen F, Mampallil D et al (2015) Chemical-free lysis and fractionation of cells by use of surface acoustic waves for sensitive protein assays. Anal Chem 87(4):2161–2169
35. Di Carlo D, Jeong KH, Lee LP (2003) Reagentless mechanical cell lysis by nanoscale barbs in microchannels for sample preparation. Lab Chip 3(4):287–291
36. Sims CE, Allbritton NL (2003) Single-cell kinase assays: opening a window onto cell behavior. Curr Opin Biotech 14(1):23–28
37. Perez-Toralla K, Mottet G, Guneri ET et al (2015) FISH in chips: turning microfluidic fluorescence in situ hybridization into a quantitative and clinically reliable molecular diagnosis tool. Lab Chip 15(3):811–822
38. Scriven PN, Ogilvie CM (2010) FISH for pre-implantation genetic diagnosis. Method Mol Biol 659:269–282
39. Dochow S, Krafft C, Neugebauer U et al (2011) Tumour cell identification by means of Raman spectroscopy in combination with optical traps and microfluidic environments. Lab Chip 11(8):1484–1490
40. El-Ali J, Gaudet S, Gunther A et al (2005) Cell stimulus and lysis in a microfluidic device with segmented gas-liquid flow. Anal Chem 77(11):3629–3636
41. Marcus JS, Anderson WF, Quake SR (2006) Microfluidic single-cell mRNA isolation and analysis. Anal Chem 78(9):3084–3089
42. Yang Y, Swennenhuis JF, Rho HS et al (2014) Parallel single cancer cell whole genome amplification using button-valve assisted mixing in nanoliter chambers. PloS One 9(9):e107958
43. Lee WC, Rigante S, Pisano AP, Kuypers FA (2010) Large-scale arrays of picolitre chambers for single-cell analysis of large cell populations. Lab Chip 10(21):2952–2958
44. Munce NR, Li J, Herman PR, Lilge L (2004) Microfabricated system for parallel single-cell capillary electrophoresis. Anal Chem 76(17):4983–4989
45. Salehi-Reyhani A, Kaplinsky J, Burgin E et al (2011) A first step towards practical single cell proteomics: a microfluidic antibody capture chip with TIRF detection. Lab Chip 11(7):1256–1261
46. Kovac JR, Voldman J (2007) Intuitive, image-based cell sorting using optofluidic cell sorting. Anal Chem 79(24):9321–9330
47. Di Carlo D, Lee LP (2006) Dynamic single-cell analysis for quantitative biology. Anal Chem 78(23):7918–7925
48. Gong Y, Ogunniyi AO, Love JC (2010) Massively parallel detection of gene expression in single cells using subnanolitre wells. Lab Chip 10(18):2334–2337
49. Salehi-Reyhani A, Sharma S, Burgin E et al (2013) Scaling advantages and constraints in miniaturized capture assays for single cell protein analysis. Lab Chip 13(11):2066–2074
50. Eyer K, Kuhn P, Hanke C et al (2012) A microchamber array for single cell isolation and analysis of intracellular biomolecules. Lab Chip 12(4):765–772
51. Brouzes E, Medkova M, Savenelli N et al (2009) Droplet microfluidic technology for single-cell high-throughput screening. P Natl Acad Sci USA 106(34):14195–14200
52. Sarkar S, Cohen N, Sabhachandani P et al (2015) Phenotypic drug profiling in droplet microfluidics for better targeting of drug-resistant tumors. Lab Chip 15(23):4441–4450
53. Phillips KS, Lai HH, Johnson E et al (2011) Continuous analysis of dye-loaded, single cells on a microfluidic chip. Lab Chip 11(7):1333–1341

Chapter 16

FISH-in-CHIPS: A Microfluidic Platform for Molecular Typing of Cancer Cells

Karla Perez-Toralla, Guillaume Mottet, Ezgi Tulukcuoglu-Guneri, Jérôme Champ, François-Clément Bidard, Jean-Yves Pierga, Jerzy Klijanienko, Irena Draskovic, Laurent Malaquin, Jean-Louis Viovy, and Stéphanie Descroix

Abstract

Microfluidics offer powerful tools for the control, manipulation, and analysis of cells, in particular for the assessment of cell malignancy or the study of cell subpopulations. However, implementing complex biological protocols on chip remains a challenge. Sample preparation is often performed off chip using multiple manually performed steps, and protocols usually include different dehydration and drying steps that are not always compatible with a microfluidic format.

Here, we report the implementation of a Fluorescence in situ Hybridization (FISH) protocol for the molecular typing of cancer cells in a simple and low-cost device. The geometry of the chip allows integrating the sample preparation steps to efficiently assess the genomic content of individual cells using a minute amount of sample. The FISH protocol can be fully automated, thus enabling its use in routine clinical practice.

Key words FISH, Gene amplification, Microfluidic, Cancer, Sample preparation

1 Introduction

The molecular characterization of cancer cells has become an essential tool in oncology as it provides relevant clinical information for treatment orientation and better patient management. Many methods are available to identify known cancer biomarkers, such as mutations in gene and protein expression levels, using cellular content from a tumor biopsy [1].

Fluorescent in situ Hybridization (FISH) is a molecular cytogenetic technique that allows determining chromosomal or genetic aberrations such as translocations, deletions, duplications, and amplifications of genes. Conventional FISH analysis consists in hybridizing a fluorescently labeled DNA probe to its complementary

Valérie Taly et al. (eds.), *Microchip Diagnostics: Methods and Protocols*, Methods in Molecular Biology, vol. 1547, DOI 10.1007/978-1-4939-6734-6_16, © Springer Science+Business Media LLC 2017

sequence in interphase nuclei or chromosomal preparations previously fixed on glass slides. Briefly, during a FISH experiment the probes and the target DNA are denatured, i.e., made single stranded by a temperature treatment. The complementary sequences of the probe and the target gene are then allowed to hybridize. After stringent washes to remove unbound and nonspecifically bound probes, fluorescent signals can be detected at the site of probe hybridization using a fluorescence microscope [2]. FISH accuracy and versatility have provided a wide range of biological and medical applications [3]. In the field of cancer research, standard FISH protocols have been useful to assess the complexity, instability, and heterogeneity of cell populations within a tumor. Indeed, in contrast to other methods based on membrane staining such as immunohistochemistry, FISH provides quantitative genetic information on a cell-by-cell basis. Finally, as it preserves cell morphology and tissue architecture, it can also be combined with further cytological and histologic analysis. Despite the advantages of this method, its use in the clinics has been limited because of its high cost and complex protocols.

Implementing a FISH protocol on chip offers several advantages such as the use of μL volumes of reagents instead of mL. In particular, DNA probes are the most expensive reagent; thus, decreasing the amount of probes required for one test can also lower its overall cost. Microfluidics also offers the possibility of creating a fully automated FISH platform to perform high-throughput analysis, from sample preparation to detection. Some examples of FISH miniaturization have demonstrated the interest of such strategies for the widespread use of FISH in routine clinical practice [4–8].

Different microfluidic-based technologies have been developed for the sample preparation of cell-based assays [9]. In particular, FISH requires proper cell immobilization on a low shear stress chamber, precise delivery of expensive probes and reagents, as well as fine control of temperature. High magnification fluorescence imaging is also crucial for accurate FISH signal analysis.

Here, we present a novel device that combines all these features using low-cost microfabrication technologies [10] (Fig. 1). To increase the potential for full automation, we have adapted and modify the standard FISH protocol (Fig. 2), as performed on glass slides, by eliminating all dehydration steps without hindering the protocol efficiency [8].

We present the protocol for the FISH analysis in microfluidic chip of ERBB2 gene amplification, a relevant clinical biomarker for breast cancer patients, and an indicator for the prescription of targeted therapies [11]. Other genetic or chromosomal abnormalities could also be assessed using this technology.

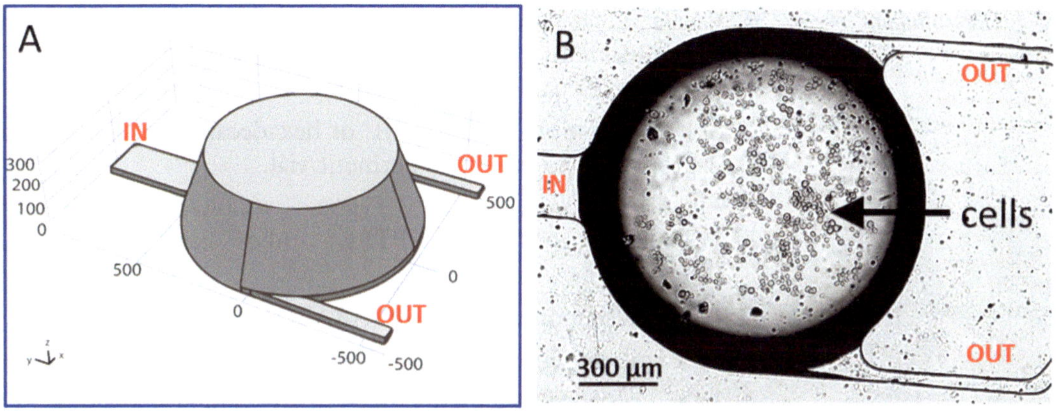

Fig. 1 (**a**) Schematic view of the "3D" chamber (height = 0.38 mm, bottom diameter = 1 mm) for FISH analysis on COC chip. Narrow microchannels (height = 30 μm, width = 60 μm) are used for the injection of cells and reagents. The difference in heights between the chamber and the channels induces a high flow velocity contrast to provide a selective immobilization of cells. (**b**) Micrograph showing the cells immobilized at the bottom of the chamber, while no cells remain on the narrow channels

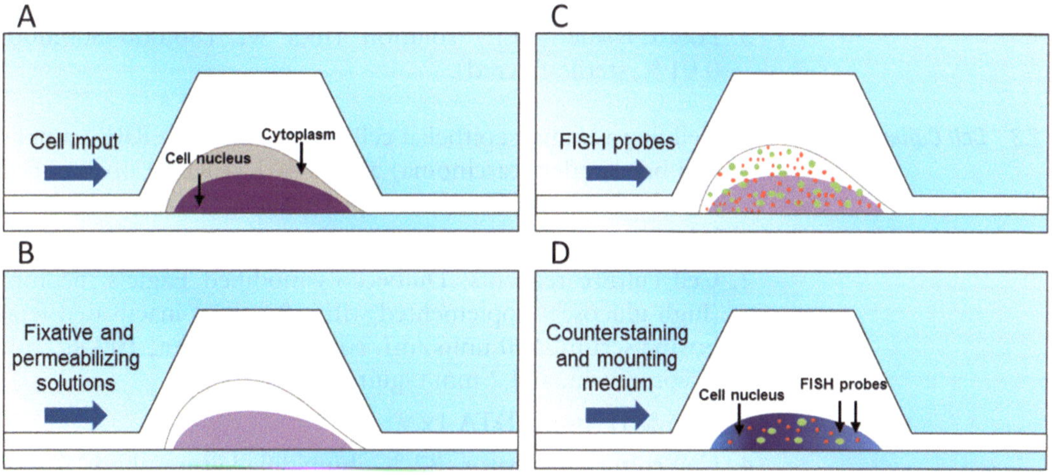

Fig. 2 Schematic representation of the different steps of the FISH protocol on COC chip: (**a**) Step1: cell immobilization on the treated COC surface, (**b**) cell fixation and permeabilization followed by a washing step, (**c**) injection of FISH probes, denaturation of probes and cell DNA, hybridization of FISH probes (*red* and *green* dots) on the target gene sequence, (**d**) Post hybridization washing to remove unbound DNA probes and counterstaining of the cell nucleus with DAPI solution (*blue*) for imaging. No dehydration and drying steps are performed in the protocol on chip

2 Materials

All solutions intended to be injected inside the microfluidic platform should be freshly prepared and filtered. All solvents and reagents should be analytical grade quality.

2.1 Microfluidic Device

1. Cyclic olefin copolymer (COC) plates (T_g = 80 °C, 5 mm thick) and films (T_g = 80 °C, 145 μm thick) (Topas® Advanced Polymers, Extrusion Lab, Germany).
2. Solvent solution: mix 1.5 mL of hexadecane and 8.5 mL of 2-propanol and store in a hermetic vial.
3. Fluidic connections: Tygon (Cole-Parmer) and Polytetrafluoroethylene (PTFE) tubes (inner diameter: 0.028 mm).
4. Glue: bicomponent adhesive (ARALDITE).
5. Hot embossing: manual hydraulic press (Specac).
6. In this protocol, we used a metal master mold obtained by 3D micro-milling (militech).
7. Hot plate (Stuart Equipment, UK) and temperature probe (Implementable Thermocouple probe, Thermo Fisher Scientific).

2.2 Surface Treatment

1. Phosphate-Buffered Saline (PBS) 1X.
2. Hydroxyethyl cellulose (HEC): prepare a fresh solution at 2 % (w/v) in PBS 1X, store at 4 °C.
3. Poly-L-lysine (PLL) solution (mol wt 150,000–300,000, 0.01 %, sterile-filtered).

2.3 Cell Culture

1. Cell lines: human epithelial cell lines SKBR-3 (ERBB2 amplified breast adenocarcinoma) and G-401 (ERBB2 unamplified rhabdoid tumor) (American Type Culture collection, Manassas, VA).
2. Cell culture reagents: Dulbecco's modified Eagle's medium (high glucose) supplemented with 10 % heat-inactivated fetal bovine serum, 100 units/mL aqueous penicillin, 100 μg/mL streptomycin, and 2 mm L-glutamine.
3. 0.05 % Trypsin-EDTA 1×.
4. Cell culture incubator (37 °C, humidified atmosphere, 5 % of carbon dioxide).
5. Centrifuge.

2.4 Fluorescence In Situ Hybridization

1. Pepsin lyophilized powder (from porcine gastric mucosa; 3200–4500 units/mg protein).

   Prepare a pepsin stock solution at 1 μg/μL and store at −20 °C.

   Freshly prepare a pepsin solution at 1 μg/mL in 0.01 M HCl from an aliquot of stock solution.
2. Fixative solution: mix 1 mL of Acetic acid and 3 mL of ethanol, store at 4 °C.
3. Sodium Saline Citrate buffer: prepare a SSC 2X solution diluted from SSC 20X, pH~7 in ultrapure water. pH can be adjusted at 7 using a 14N HCl solution.

4. Washing solution I: 0.4X SSC +0.3 % NP40, pH~7.

   Mix 10 mL of SSC 2X solution with 150 μL NP40 at room temperature and add 40 mL of ultrapure water. Adjust the pH at 7 with HCl if necessary.

   Preheat 5 mL of washing solution I for 30 min at 72 °C prior to use.

5. Washing Solution II: 2X SSC +0.1 % NP40, pH~7.

   Mix 10 mL of SSC 2X solution with 100 μL NP40 at room temperature and add 90 mL of SSC 2X. Adjust the pH at 7 with HCl if necessary.

6. DAPI (4′,6-diamidino-2-phenylindole) counterstain (0.2 μg/mL) and Vectashield.

7. Ready-to-use FISH probes (ON ERBB2, HER2/*neu* (17q12) /SE17) and FISH Hybridization Buffer (FHB) (Kreatech Diagnostics, The Netherlands).

8. Optional: Prepare a RNAse stock solution at 10 mg/mL in SSC2X and store at −20 °C.

   Freshly prepare a RNAse solution at 0.1 mg/mL from an aliquot of stock solution.

2.5 Fluorescence Imaging

1. Epi-fluorescence microscope (Nikon Eclipse 80i) with a mercury lamp (HGFIL Lampe 130 W).
2. Piezo focus lens positioner (P-721 PIFOC®, Germany) or automated stage for fine *z* stacking and (*x*, *y*) positioning.
3. High-speed CCD camera (CoolSNAP HQ2, Photometrics, Roper scientific–Princeton instruments).
4. Adequate filter sets (DAPI, GFP, RHOD).
5. Oil immersion 100X objective (Nikon, CFI Plan Apo VC, NA 1.4, WD 0.13).
6. Imaging Software (MetaMorph®, Molecular Devices) (*see* **Note 1**).

3 Methods

3.1 Microfluidic Device Design and Fabrication

1. Cut the COC plates in a 2.5 × 3.5 cm size, clean the plates and the template with 2-propanol, and dry under pressurized air.
2. Preheat the hot press at 130 °C. The upper plate of the hot press should not be heated to avoid deformation of the COC.
3. Place COC substrate on the top of the master, load into the press, and emboss the plates at 50 kPa of pressure, 130 °C for 10 min (*see* **Note 2**).
4. Cool down the press to 40 °C at 50 kPa, unload the COC substrate and the master, and separate both parts.

5. Trim and drill the embossed COC substrate to create access holes. The size of the access holes should match the outer diameter of the capillary tubes (*see* **Note 3**).
6. Clean the chips in 2-propanol in an ultrasonic bath for 10 min and dry.
7. Apply the solvent mixture on the surface of the COC substrate containing the channels by means of a lint-free wiper. Place a COC film on the treated surface and manually apply low pressure to ensure contact.
8. Load the chip into the press at 60 °C for 10 min.
9. Place the tubes in the inlets and outlets to create fluidic connections and add glue around the tubes to ensure stability (*see* **Note 4**).
10. Couple the COC chip with a flow control system (MFCS, Fluigent) for the injection of all reagents (*see* **Note 5**).

3.2 Surface Treatment

1. Inject 200 μL of HEC solution in the microchannels and let to incubate for 1 h at room temperature (RT).
2. Inject 200 μL of PBS as a rinsing step.
3. Inject 100 μL of PLL and let to incubate for 3 h at 37 °C on the top of a hot plate.
4. Inject 200 μL of PBS as a final rinsing step before injecting the sample.

3.3 Cell Culture and Sample Preparation

The human epithelial cell lines SKBR-3 (*ERBB2* amplified breast adenocarcinoma) and G-401 (*ERBB2* unamplified rhabdoid tumor) were obtained from the American Type Culture collection (Manassas, VA). Cells were cultured in Dulbecco's modified Eagle's medium (high glucose) supplemented with 10 % heat-inactivated fetal bovine serum, 100 units/mL aqueous penicillin, 100 μg/mL streptomycin, and 2 mm L-glutamine, at 37 °C in a humidified atmosphere with 5 % of carbon dioxide.

1. Detach the confluent culture cell lines using a trypsin treatment (0.05 % Trypsin-EDTA 1×).
2. After brief centrifugation, resuspend the cells in 1 mL culture medium to obtain an approximate concentration of 10^6 cells/mL.
3. Inject 5–20 μL of cell suspension in the COC chip.
4. When all chambers contain cells in solution, stop the flow rate and let to incubate for 5 min at RT or 37 °C. Cells will adhere to the COC surface via the surface treatment.
5. Inject 100–200 μL of PBS 1X to remove all cells that did not adhere to the surface (*see* **Note 6**).

3.4 FISH Protocol on Chip

FISH experiments on cells adhered to glass slides should be used as controls. Here, only the protocol for FISH on chip is presented.

1. Inject 200 μL of SSC 2X buffer.
2. Inject 100 μL of pepsin solution and incubate for 10 min at 37 °C.
3. Inject 150 μL of SSC 2X buffer as a washing step.
4. Inject 100 μL of fixative solution and incubate at 4 °C for 20 min.
5. Inject 200 μL of SSC 2X buffer as a washing step and incubate at 37 °C for 10 min.
6. Perform a second washing step with 200 μL of SS2X buffer to completely remove the fixative solution (*see* **Note 7**).
7. Inject 10 μL of FISH probes and place the chip on a hot plate at 75 °C for 15 min. A very viscous solution can be injected at a maximum flow rate (*see* **Note 8**).
8. Incubate at 37 °C overnight in a humid atmosphere.
9. Inject 100 μL of washing solution 1 and incubate at 72 °C for 4 min.
10. Inject 100 μL of washing solution 2 at RT.
11. Inject 10 μL of DAPI diluted in antifading mounting medium.
12. Close the inlets and outlets and store the chip at 4 °C for further analysis.

3.5 Image Acquisition and Treatment

1. Place the chip on the stage of the fluorescence microscope and adjust the focus at the desired location using low magnification and DAPI filter set. The area should contain a high density of intact nonoverlapping nuclei.
2. Once the target cells are located, change for the 100X objective and the GFP and RHOD filter sets. Visually inspect that FISH signals are bright and homogeneous for all imaged cells and set the acquisition parameters accordingly (*see* **Note 9**).
3. Set a 0.2 μm (z) step in the piezo scanner and scan the target cells so as to capture all FISH signals (5–20 μm length) (*see* **Note 10**).
4. Repeat the imaging process at different locations of the chip. At least 30 cells should be imaged per experiment.
5. The acquired "z-stacks" can be analyzed directly or processed using the "Meinel" algorithm for 3D deconvolution [12].

3.6 FISH Analysis and HER2 Scoring Criteria

The acquired images can be further processed using ImageJ software (http://rsb.info.nih.gov/ij) as follows:

1. Create a z projection for each "z-stack" using the "Maximum intensity" criteria, for DAPI, GFP, and RHOD images independently.

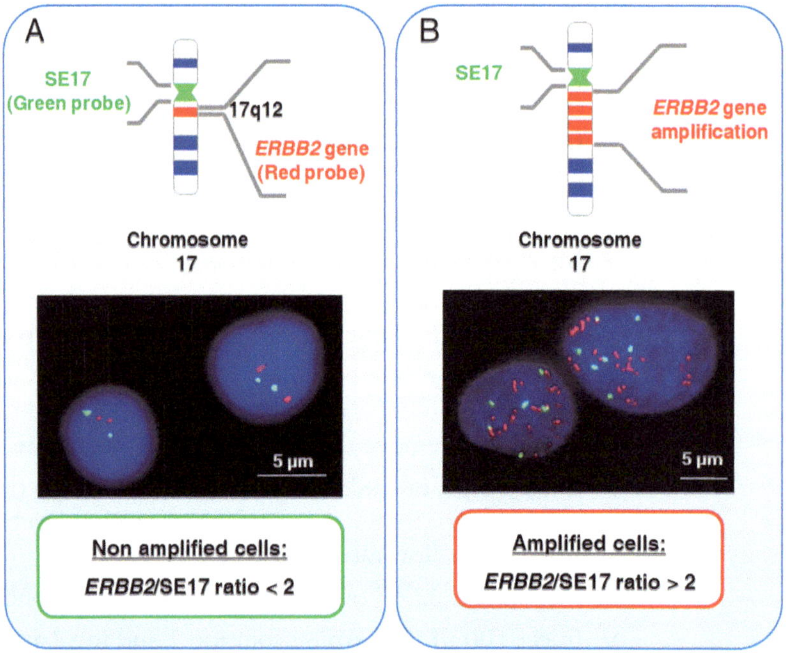

Fig. 3 Schematic representation of the location of the ERBB2 gene (**a**) and its amplification on chromosome 17(**b**). High magnification fluorescence imaging of FISH results for G-401 control cells (**a**) and for SKBR3 amplified cells (**b**)

2. Use the "color combine" tool to create a multichannel image (DAPI = blue, GFP = green, RHOD = red).
3. Adjust the brightness and contrast of each channel independently, as desired, avoiding intensity saturation.

The processed images can be analyzed as follows (Fig. 3):

1. Identify intact and nonoverlapping nuclei, showing clear and distinct red and green FISH signals.
2. Count the number of green and red signals per cell for at least 30 cells.
3. Evaluate the ERBB2 amplification score by calculating the ratio of red ERBB2 signals to green SE17 signals and apply the following criteria:
   a. Positive gene amplification: *ERBB2*/SE17 ratio ≥ 2.0 with an average *ERBB2* copy number ≥ 4.0 signals.
   b. Negative gene amplification: *ERBB2*/SE17 ratio <2.0 with an average *ERBB2* copy number <4.0 signals.
   c. Positive gene amplification with SE17 polysomy: If copy SE17 number >2 and *ERBB2* copy number ≥6.0 signals.
   d. In case of doubt, analyze 30 supplementary cells or repeat the experiment. Please refer to the 2013 ASCO/CAP guidelines [13].

4 Notes

1. Imaging can be performed manually if no automated microscope is available.
2. Embossing parameters should be adjusted when using different equipment or COC substrates with different Tg.
3. We perform a two-step procedure to create access holes; first on the front side, using a 0.5 mm drill bit to make an inlet as small as possible. Then on the back side of the substrate, a 3 mm deep hole is made using a 1.2 mm drill bit to introduce the capillaries.
4. Using glue is a simple and effective method to secure tubes in place but other methods can also be used, e.g., upchurch frits.
5. Reagents can also be injected manually by pipetting or using syringe pumps.
6. Keeping track of the regions with high density of cell in each of the chip chambers will shorten the imaging step.
7. Critical step, no fixative solution should remain on the channels prior to injecting FISH probe solution.
8. The concentration of FISH probe solution can be decreased from 1× to 0.25× by dilution in FHB.
9. Acquisition parameters can differ for each of the filter sets, but should remain constant during the complete imaging process. Use the intensity histogram to avoid signal saturation.
10. Cells on chip often show a more round configuration as compared to glass adhered to glass slides; therefore, the scan length should be increased

Acknowledgments

This work was supported by EU funding (FP7 Diatools European Projects and CellO ERC advanced grant).

References

1. Yu M, Stott S, Toner M et al (2011) Circulating tumor cells: approaches to isolation and characterization. J Cell Biol 192:373–382
2. Gole L (2001) Fluorescence in situ hybridization. Encyplodedia life Sci. doi:10.1038/npg.els.0001468
3. Levsky JM, Singer RH (2003) Fluorescence in situ hybridization: past, present and future. J Cell Sci 116:2833–2838
4. Sieben VJ, Debes-Marun CS, Pilarski LM, Backhouse CJ (2008) An integrated microfluidic chip for chromosome enumeration using fluorescence in situ hybridization. Lab Chip 8:2151–2156
5. Zanardi A, Bandiera D, Bertolini F et al (2010) Miniaturized FISH for screening of onco-hematological malignancies. Biotechniques 49:497–504

6. Vedarethinam I, Shah P, Dimaki M et al (2010) Metaphase FISH on a chip: miniaturized microfluidic device for fluorescence in situ hybridization. Sensors 10:9831–9846
7. Tai C-H, Ho C-L, Chen Y-L et al (2013) A novel integrated microfluidic platform to perform fluorescence in situ hybridization for chromosomal analysis. Microfluid Nanofluidics 15:745–752
8. Perez-Toralla K, Mottet G, Guneri ET et al (2015) FISH in chips: turning microfluidic fluorescenc-e in situ hybridization into a quantitative and clinically reliable molecular diagnosis tool. Lab Chip 15:811–822
9. Mach AJ, Adeyiga OB, Di Carlo D (2013) Microfluidic sample preparation for diagnostic cytopathology. Lab Chip 13:1011–1026
10. Mottet G, Perez-Toralla K, Tulukcuoglu E et al (2014) A three dimensional thermoplastic microfluidic chip for robust cell capture and high resolution imaging. Biomicrofluidics 8:024109
11. Arteaga CL, Sliwkowski MX, Osborne CK et al (2012) Treatment of HER2-positive breast cancer: current status and future perspectives. Nat Rev Clin Oncol 9:16–32
12. Sibarita J (2005) Deconvolution microscopy. Adv Biochem Eng Biotechnol 95:201–243
13. Wolff AC, Hammond MEH, Hicks DG et al (2013) Recommendations for human epidermal growth factor receptor 2 testing in breast cancer: American Society of Clinical Oncology/College of American Pathologists clinical practice guideline update. J Clin Oncol 31:3997–4013

INDEX

A

Antibody ... 26, 37–47, 49–54, 59, 62, 69, 71, 72, 74, 77, 79, 115, 168, 170, 171, 173, 179, 185, 198, 206, 207
Antibody-immobilized microbeads ... 49–51, 53, 54
Antigens ... 25, 45, 46, 49, 50, 53, 69, 77, 168, 170

B

Bacteremia ... 188
Bacteria ... 175–185, 188
Bacterial blood stream infection (BSI) ... 175, 176
β-cells ... 58, 66
β-Galactosidase ... 87, 88, 93, 95
Biomarker ... 15, 77, 114, 133, 143, 167, 188, 212
  protein markers ... 50, 70, 211
  nucleic acid markers
Biotinylated β-Galactosidase (BβG) ... 88, 93, 95, 96, 99
Blood ... 58, 70, 143, 167–185, 188, 197
  cells ... 168, 176, 177, 181, 182, 184, 185, 188
  stream infection ... 175, 176, 178–185

C

Cancer
  cell ... 114, 143, 145, 169, 173, 188, 190, 191, 197, 198, 206, 211–219
  cell isolation ... 167, 169, 172, 173, 188
  circulating tumor cells (CTC) ... 167–174, 188, 197, 204
  colon cancer ... 167
  lung cancer ... 167, 169
  metastasis ... 167
  mutations ... 143, 168, 211
Capillary-driven flow ... 25–47
Capillary electrophoresis (CE) ... 206
Cell
  β-cells ... 58, 66
  fluorescence in situ hybridization (FISH) ... 206, 211–219
  lysis ... 176–178, 181–182, 184, 191, 193, 195, 198–201, 205–207
  negative enrichment of circulating tumor cells (CTCs) ... 167–174
  permeabilization ... 191, 193, 198–202, 205, 206, 213
  single cell ... 57, 114, 187–207
  sorting
  staining ... 171–173, 190–191, 195, 197–198, 206
  trapping ... 189, 190, 192, 193, 195–204, 206, 207
  white blood cell (WBC) depletion ... 176
Chip singulation ... 30–34
Clinical diagnostics ... 5, 37, 69–81
Commercialization ... 5, 6
Competitive immunoassay ... 58
Continuous flow ... 70
C-reactive protein (CRP) ... 50, 53–55, 70–72, 74, 76–79

D

Deflection ... 70
Diabetes ... 58
Diagnostics ... 3–18, 85, 144, 175–176, 187–207
  cartridge ... 8, 10, 12–16, 18
Digital bioassays ... 99
Digital microfluidics (DMF). *See* Microfluidics
Digital PCR ... 124, 125, 128, 129, 143–162
DNA
  amplification ... 140, 144, 145, 156, 160
  detection ... 87, 144, 145
  digital PCR ... 124, 125, 129, 143–162
  gene amplification ... 158, 211, 212, 218
  multiplexing ... 133, 134
  mutations ... 143–145
  nucleic acid amplification ... 8, 123–131
  PCR ... 134, 144, 145, 149, 154, 156
  probes ... 37, 145, 154, 156, 211–213
  single-molecule detection ... 85–100, 105–110
  stretching ... 105–110
  trapping ... 106–109
Dry-film resist (DFR) ... 26–47

E

Electrophoresis ... 8, 58–60, 62, 64, 189, 206
Electrowetting ... 86, 97
Electrowetting-on-dielectric-based digital microfluidics (EWOD-DMF) ... 86

Valérie Taly et al. (eds.), *Microchip Diagnostics: Methods and Protocols*, Methods in Molecular Biology, vol. 1547, DOI 10.1007/978-1-4939-6734-6, © Springer Science+Business Media LLC 2017

Epithelial cell adhesion molecule (EpCAM) 191, 197, 198
Escherichia coli 94
Eucaryotic cell

F

Femtoliter-sized microwells 85–100
Fluorescence in situ hybridization (FISH). *See* Cell

G

Gene amplification 158, 212, 218
Glucose detection
Glucose sensitivity 58, 63, 66

H

Hoechst 33342 170, 172, 173, 191
Hormone monitoring 57–66
Hydrogel 49–54

I

Immunoassays 14–17, 25–47, 49–54, 57–66, 69–81, 125
Immunocytochemistry 169
Immunomagnetic separation 169, 172
Immuno-pillars device 50, 51, 53, 54
In vitro diagnostics 58, 197, 204
Infections 17, 70, 175–185
Inflammation 70
Islets of Langerhans 58

K

KRAS 133–141, 143–162

L

Lab-on-a-chip 7, 34
Laminar flow 27, 40, 70, 71, 77, 81
Laser induced fluorescence (LIF) detection 62–63

M

Magnetic particles 16, 69–81, 85–100, 113–120, 169, 171
Magnetic tweezers 114, 116, 117, 119
Marker 90
Microbeads 49–54
Microcontact printing 26, 37–47
Microfabrication 26, 38, 106, 114, 212
Microfluidics
  chip fabrication 59–60, 215–216
  device 6, 8–11, 25, 37, 38, 49, 57, 58, 61, 69, 72, 75, 79, 80, 106, 114, 141, 149, 156, 162, 176, 177, 180–181, 189, 190, 192–199, 201, 214–216
  digital microfluidics 85–100
  droplet-based 113–120, 135, 143–162, 207
  immunoassay 49–54
  lab-on-a-chip 7, 34
  silicon–based microfluidic chips 7, 13, 15, 26
Microparticles 15, 173
Micropillars 49–54, 70
Micro slit membrane 168, 170, 171, 173
Microtas, μ-TAS 3, 6
Microwell array 86, 87, 89, 91, 93, 95, 96, 98, 99, 207
Molecular diagnosis 8, 12, 14–16, 18, 129, 176
Molecular patterning 105, 106
Multilaminar flow 69–81
Multiplexing 133–141

N

Nucleic acid amplification 8, 123–131

O

Optical mapping 105

P

Parallel single cell analysis (PaSCAl) 187–207
Parylene filter membrane 169
Passive droplet fusion 133–141
Pathogens 25, 176
Patterning antibodies 26, 37–47
Perfusion 57, 59–61, 64
  culture 59
  reservoir 61
Photomasks 26, 28, 29, 33, 34, 39, 51, 53, 54, 59, 64, 72, 89–91, 95, 106, 125, 126, 136, 148, 178, 179
Picoliter size droplets 153
Point-of-care diagnostics 16, 18, 38, 113
PolydimethylSiloxane (PDMS) 25, 35, 38, 39, 41–46, 106, 108–110, 116–118, 120, 135–137, 146–148, 150, 176–178, 180–183, 189, 192, 194, 203, 206, 207
Polymerase chain reaction (PCR) 8, 12, 14, 16, 114, 115, 118, 119, 124–126, 128, 129, 134–137, 139–141, 143–162, 176, 206, 207
Polymethyl methacrylate (PMMA) 168
Probes 37, 92, 116, 131, 134–141, 145–147, 154, 156, 157, 161, 191, 195, 198, 211–215, 217, 219
Product development 3, 4, 7–15, 18

Q

Quantification 86, 96, 123, 124, 129, 139, 144, 145

R

Reverse transcription (RT) 114, 115, 117, 119, 145

S

Sample preparation....6, 8, 171, 176, 190, 194–195, 212, 216–217
Sandwich immunoassays.... 15, 46, 70, 71
Sealing....11, 25–36, 38, 40, 44–46, 85–100, 126, 196, 204
Sepsis....175, 177
Silicon–based microfluidic chips....7, 13, 15, 26, 194
Single cells.... 57, 114, 187–207
Single-molecule detection....85–100
Soft-lithography....85–100, 105–110, 116, 120, 123, 136, 144, 148, 176, 194, 207
Streptavidin-coated superparamagnetic particles....72, 79, 88, 94, 97
Superparamagnetic particles....72, 79, 88, 93, 94
Surfactant....88, 116, 118, 135, 137, 138, 146, 149

U

Ultrasensitive detection....123